Pinewood Derby® Speed Secrets

Tell me, and I'll forget. Show me, and I may not remember. Involve me, and I'll understand.

—*Native American Proverb*

Pinewood Derby® Speed Secrets

by David Meade
with illustrations
by Troy Thorne

Fox
Chapel Publishing

1970 Broad Street • East Petersburg, PA 17520
www.FoxChapelPublishing.com

Alan Giagnocavo
Publisher

Peg Couch
Acquisition Editor

Gretchen Bacon
Editor

Troy Thorne
Design and Layout

Greg Heisey
Photography

Note to Authors: We are always looking for talented authors to write new books in our area of woodworking, design, and related crafts. Please send a brief letter describing your idea to Peg Couch, Acquisition Editor, 1970 Broad Street, East Petersburg, PA 17520.

To learn more about the other great books from Fox Chapel Publishing, or to find a retailer near you, call toll-free 1-800-457-9112 or visit us at www.FoxChapelPublishing.com.

Pinewood Derby®, Cub Scouts®, Regatta®, Space Derby®, and Boy Scouts of America® are registered trademarks of the Boy Scouts of America. Printed under license from Boy Scouts of America to DK Publishing.

ISBN-13: 978-1-56523-291-4
ISBN-10: 1-56523-291-7

Publisher's Cataloging-in-Publication Data

Meade, David.
    Pinewood Derby speed secrets / by David Meade ; with illustrations
    by Troy Thorne. -- East Petersburg, PA : Fox Chapel Publishing,
    2006.

        p. ; cm.

        ISBN-10: 0-7566-2733-8 (DK)
        ISBN-13: 978-0-7566-2733-1 (DK)
        ISBN-10: 1-56523-291-7 (Fox Chapel)
        ISBN-13: 978-1-56523-291-4 (Fox Chapel)

        Includes photographs, instructional diagrams, design templates,
    a gallery of cars, and tips to help you and your child gain the
    competitive edge in a Pinewood Derby race.

        1. Model car racing. 2. Automobiles--Models. 3. Automobiles--
    Design and construction. I. Thorne, Troy. II. Title.

GV1570 .M43 2006

796.15/6--dc22   0610

Printed in China
10 9 8 7 6 5 4 3 2

A Note from the Author

The night of my oldest son's first Pinewood Derby® was one of the most memorable events of my life. Neither of us knew exactly what to expect. We had worked together to incorporate all of the steps collected in this book, steps that are part of our family's tradition of unbeatable cars. I wish that I somehow could have captured the excitement on his face as his car won race after race. He was so proud when he was awarded the first-place trophy for the entire pack.

Winning the Pinewood Derby was exciting and fun; however, more important than the outcome was the Derby's positive effect on our relationship. Remember, even though I will spend a lot of time in this book talking about principles that will help you win, the Pinewood Derby is supposed to be fun. It's supposed to be exciting! Its purpose is to help children develop skills, confidence, and self-esteem. Don't forget that your Derby experience is only a success if you enjoy the ride.

For my family, building Pinewood Derby cars that win and having a fun time doing it is a tradition that goes back generations. Some of my most vivid and cherished memories are of the time I spent crafting Derby cars with my father and brothers. Now that I have sons of my own, it has been great to see them build identity and confidence and take part in a family tradition.

By using our family's techniques—the information provided in this book—my sons have taken first place in every Pinewood Derby we have raced in. We have raced in four packs in Texas and Wisconsin, and some of the Derbies have sported over 100 cars. Many people have shown up with cars they built using advice they found on the Internet, but our cars still come in first place.

Now it's your turn. By following the steps in this book, you'll get our tried-and-true techniques that make this the most complete guide to building a winning car on the market today. This is not just another speed tips book. This book will provide you with step-by-step instructions, photos, and diagrams for the entire process so that you and your child can have fun, build a stronger relationship, and go home winners.

David Meade got his start in the Pinewood Derby as a boy, over 20 years ago, when his family was well known for building unbeatable Pinewood Derby cars. Today, David's own sons carry on the winning tradition and have garnered 13 first-place championships for seven consecutive years.

In addition to his involvement in his sons' Pinewood Derby activities, David runs *www.pinederbyinfo.com*. Because of his success, David was interviewed by the *Wall Street Journal* and was quoted in "Driving Under the Influence—of Dad" for the March 4, 2005 issue.

By profession, David is a research scientist, holding a master's degree in statistics. His background in research and experimental design has enabled him to develop a unique and powerful approach to Pinewood Derby racing. He also enjoys model rocketry and astronomy.

Table of Contents

Pinewood Derby strengthens relationships, while promoting craftsmanship and healthy competition

The moment the first group of miniature cars started down the 32-foot race ramp with the battery-run finish line made from doorbells, the Pinewood Derby enjoyed instant success.

The brainchild of California Cubmaster Donald Murphy, the Derby arose from his search for an activity that he and his 10-year-old son could work on together. Murphy, an art director, was inspired by his employer, North American Aviation, which sponsored Soap Box Derby races, as well as by his own childhood experiences.

"I'd made models of airplanes, cars, boats, and any number of other structures and remembered the pleasure I got out of doing it," Murphy told *Scouting* magazine in November 1999. " I also wanted to devise a wholesome, constructive activity that would foster a closer father-son relationship and promote craftsmanship and good sportsmanship through competition."

He presented his idea for carving and racing miniature cars to Cub Scout Pack 280C of Manhattan Beach, California, which heartily endorsed the project. He then proceeded to design a gravity-propelled car that could be carved out of soft pine and wrote the rules for racing the miniature vehicles. Those rules stated, "The Derby is run in heats on a two- to four-lane track. Two to four cars starting from a standstill will run self-propelled down an inclined track to the finish line. Cars are guided by a raised spacer on the track between the wheels."

The first Pinewood Derby was held May 15, 1953, at the pack's newly constructed Scout House in Manhattan Beach. The Management Club at North American Aviation sponsored the event. Contestants from the 55-member Cub Scout pack, using kits consisting of a block of pine, two wooden axles, four nails, and four wheels, raced in three classes: Class A, for 10-year-olds; Class B, for 9-year-olds; and Class C, for 8-year-olds.

The following year, a local newspaper and the Los Angeles City Recreation and Parks Department sponsored a citywide Pinewood Derby, and the national office of the Boy Scouts of America® adopted the program for Cub Scout packs nationwide.

More than a half century after the running of the first Pinewood Derby, the event continues to be immensely popular not only among Cub Scouts® (more than 100 million of which have participated in the Derby to date), but with members of other youth groups as well. The program has been adopted by the Girl Scouts®, Awana® Clubs International (as the Awana Grand Prix), Scouts Canada (as the Kub Kar Rally), the Christian Service Brigade® (as the Shape N Race Derby), the Royal Rangers®, the YMCA Adventure Guides®, and the Woodcar Independent Racing League (WIRL). It also has spawned related events, such as the Raingutter Regatta®, which uses boats instead of cars, and the Space Derby®, which uses rockets.

This car, raced in the first Pinewood Derby in 1953, features a pair of engine intake stacks on its hood and a wire exhaust pipe along the side.

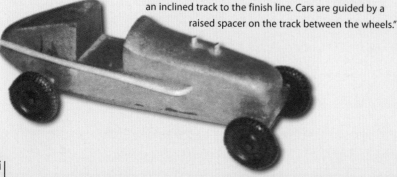

Race day, 1953. The first Pinewood Derby was such a huge success that the Boy Scouts of America adopted the program nationwide the following year.

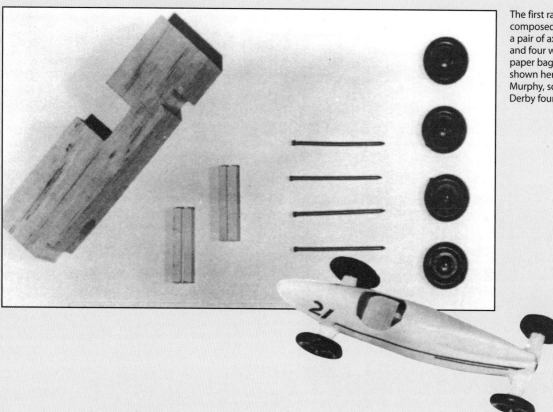

The first racing kits were composed of a block of wood, a pair of axles, four nails, and four wheels packed in a paper bag. The finished car shown here belonged to Donn Murphy, son of Pinewood Derby founder Don Murphy.

The Pinewood Derby has also spawned a whole genre of products and services for car builders and racers.

When the Pinewood Derby observed its fiftieth birthday in 2003, Murphy, then 83 years old, was honored with a proclamation from the president of the United States. He also received commendations from the national director of Cub Scouting, the governor of California, both of California's U.S. senators, and the mayor of Los Angeles. In Manhattan Beach, the Derby's birthplace,

Pack 713, a direct descendant of Pack 280C, hosted a commemoration for the city's Cub Scouts, staging several races with retro 1953 cars in the same Scout House that was the site of the first Derby.

Because of Don Murphy, father of the Pinewood Derby, generations of parents and children have built strong relationships as they developed valuable skills.

His legacy lives on.

Winners of the first Pinewood Derby in 1953, from left: Dennis Lowrie, Class C, age 8; Richard Hoffman, Class A, age 10; and Mike Thorne, Class B, age 9. Hoffman also won the grand prize trophy for classes B and C. Pack institutional representative Jack Swofford is pictured at right.

car plans

NOTES:

A) CAR MUST BE MADE TO DIMENSION AS NOTED TO FIT RACING TRACK.

B) DESIGN IS OPTIONAL.

C) ASSEMBLE PARTS WITH MODEL CEMENT.

D) USE WOOD FILLER FOR FAIRING.

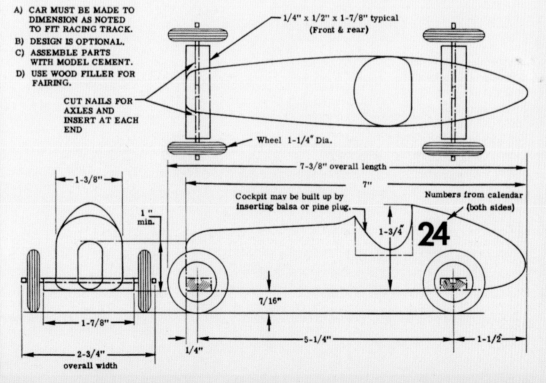

1/4" x 1/2" x 1-7/8" typical (Front & rear)

CUT NAILS FOR AXLES AND INSERT AT EACH END

Wheel 1-1/4" Dia.

7-3/8" overall length

7"

1-3/8"

1" min.

Cockpit may be built up by inserting balsa or pine plug.

1-3/4"

Numbers from calendar (both sides)

24

7/16"

1-7/8"

2-3/4" overall width

1/4"

5-1/4"

1-1/2"

Original car plans from the 1953 official rule book. While the plans remain basically the same, car designs have changed with the times and the whims of their owners.

I have designed this book to be useful to nearly everyone, regardless of experience. If this is your very first Pinewood Derby, that's okay. I'll take you by the hand and lead you down the path as far as you feel comfortable. If you're experienced, this book will teach you the advanced techniques you need to win.

Everyone knows the basic steps of building a Pinewood Derby car—get the kit, cut the block, glue the wheels, and show up on race day. But winners know that there's much more to it. Building a winning car is not hard, but it does require that you do more than the basic steps. And there's no one big solution. I wish I had a dollar for every person who has come up to me after a race, looked at our car, and then said, "So what magic thing did you do to this car?" Unfortunately, there is no magic step or secret. Winners follow a series of small steps, each of which will shave fractions of seconds off your race time. These fractions add up. Combined, they'll save you precious seconds and allow you to take home the trophy.

This process of taking small steps is the reason I've designed this book differently. I could have taught you step-by-step how to build a car, but that would be boring. Everyone would show up with the same car on race day. And some of the techniques may or may not be legal for your Derby. (Remember, rules and regulations vary from pack to pack; check your local guidelines before proceeding with any suggestion contained in this book.)

Instead, I'm giving you the information you'll need in three main chapters for the body, the axles, and the wheels. Within each of these large sections, I've organized all of the techniques into three levels: Winning (beginning), Champion (intermediate), and Ultimate (advanced). I've also included a Getting Started chapter with some general information on safety, getting your child involved, principles of speed, and car design and a final chapter that will get your car ready for the day of the race.

This format allows you to:

• Learn all of the techniques.

• Choose the tips and techniques that fit your skill level, use the tools you have, and comply with your Derby's rules.

• Build on your skills each year you participate in the Derby.

• Get right down to the parts that are new for each level, rather than reading through the same steps for each level of car.

No matter what level or tips you choose, you'll end up with a competitive car that you and your child can be proud of.

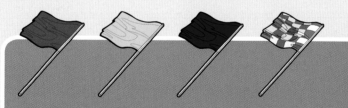

Tip Boxes with Flags

The tips throughout the book feature multicolored flags to help you along the way. Red flags indicate processes or items to avoid, yellow flags denote safety information or precautions that should be taken, green flags show alternate ways to accomplish tasks or further information, and checkered flags point to speed tips or speed theory.

Now, before you jump right in, it's important that you read this book in its entirety. Remember, you want to get an overview of all the processes, and then select the steps that are right for you. Here's more to help you navigate through the book:

Tabs and bars: The color-coded tabs on the sides of the pages and bars at the tops of the pages indicate which level is covered on that page. Red denotes the Winning Car (beginning), blue denotes the Champion Car (intermediate), and yellow denotes the Ultimate Car (advanced). The gray sections apply to all cars.

Diagram of steps: This illustration is your visual map to completing the steps for a particular level (Winning, Champion, Ultimate) within the chapters. Be sure to follow the order carefully. Since each car builds on the one before, all steps may not be provided to avoid duplication. Page numbers in each annotation will point you to the appropriate demonstration.

Materials and tools lists: Within each chapter (body, axles, wheels), a complete list of materials and tools is provided for each level (Winning, Champion, Ultimate).

So, hang on tight and get ready to learn how we build cars that leave all the others in the dust!

Pages with a gray bar and multicolored tabs apply to all cars.

Pages with red bars and tabs apply to the Winning Car (beginning).

Pages with blue bars and tabs apply to the Champion Car (intermediate).

Pages with yellow bars and tabs apply to the Ultimate Car (advanced).

Materials and tools lists found at the beginning of each section give you all of the items needed for the level within the chapter.

Follow the diagram of steps carefully to complete your chosen tips in the proper order.

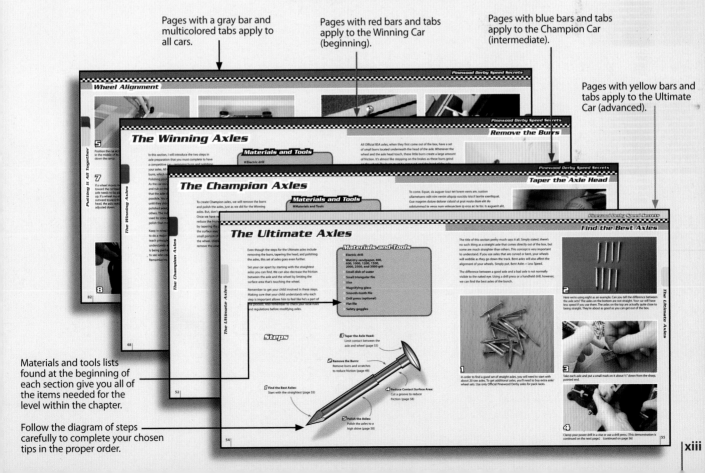

Chapter 1

Getting Started in the Pinewood Derby

Whether you are new to the Pinewood Derby or have been involved for a number of years, take the time to read over this section. Having a solid understanding of the principles of safety, speed, and car design is essential to building your car and will make the process more fun for you and your child.

Safety Essentials. A dust mask, goggles, and latex gloves are indispensable safety tools for building a Pinewood Derby car.

The Pinewood Derby is all about having fun; however, the fun quickly disappears when someone gets hurt. Make safety your top priority. The list below outlines some general safety rules that should be followed.

Wear eye protection at all times. You only have one set of eyes. Keep them safe. A good pair of safety glasses or goggles that protects your eyes from the front and the sides is ideal.

Monitor your child's use of tools. Children should not use power tools without adult supervision. Some power tools, such as an electric miter saw or a router, should not be used by children at all.

Wear a dust mask when appropriate. Sanding, applying graphite, and spray painting are examples of activities that require the use of a dust mask to protect against harmful fumes and airborne particles. A disposable dust mask should work well.

Work in a well-lighted and well-ventilated area, especially when performing tasks that involve fumes or airborne particles.

Consider wearing gloves when using sharp tools. Leather gloves that fit snugly are a good choice.

Do not wear loose-fitting clothing when operating power tools. Loose clothing can become caught or snagged by moving parts.

Do not melt lead. Molten lead is extremely hot and very poisonous. There are numerous and better ways to safely add weight to your car without melting lead.

Melting or Sanding Lead

There are a lot of Pinewood Derby websites and "How to Win" books and videos out there, several of which actually recommend melting or sanding lead. These techniques are very dangerous and will not help you win the Derby. Remember, lead is very poisonous, and, when you melt it or sand it, you create lead dust and fumes that can easily be breathed into your lungs.

Handle lead with care. There are appropriate ways to add weight to your car using lead. Whenever you are handling lead, always wear latex gloves. If you do touch lead with your bare hands, be sure to wash your hands thoroughly afterward. Keep all lead products out of the reach of children. All forms of lead are very poisonous!

Follow all safety rules and precautions listed on the tools and products you use during the construction process.

Keep your work area clean and organized. A workspace that is messy and cluttered is an accident waiting to happen. Tools should be organized and kept in safe locations, and the floor and workspace should be free of clutter. Make sure power cords don't present a tripping hazard when you are walking around them.

This principle from the Boy Scout Law applies to everyone. Honesty is the true measure of any human being. When you win the Pinewood Derby, do so honestly. Follow your local rules. If one of the speed tips discussed in this book is not allowed in your Derby, don't use it. If you want to use a particular speed tip but you aren't sure whether it's legal, clear it with your local race organizer before you start building your car. Be honest— it's the Scouting way!

At the right is a list of the suggested rules that accompanies each Official Grand Prix Pinewood Derby Kit®. Ask your local race committee for a copy of your rules and then abide by them.

Official Grand Prix Pinewood Derby Rules

1. Wheel bearings and bushings are prohibited.

2. The car shall not ride on springs.

3. Only official Cub Scout Grand Prix Pinewood Derby wheels and axles are permitted.

4. Only dry lubricant is permitted.

5. Details, such as steering wheel and driver, are permissible as long as these details do not exceed the maximum length, width, and weight specifications.

6. The car must be free-wheeling, with no starting devices.

7. Each car must pass inspection by the official inspection committee before it may compete. If, at registration, a car does not pass inspection, the owner will be informed of the reason for failure and will be given time within the official weigh-in time period to make the adjustment. After approval, cars will not be re-inspected unless the car is damaged in handling or in a race.

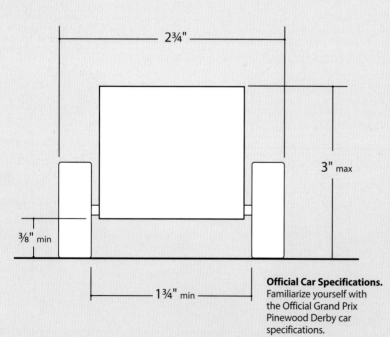

Official Car Specifications. Familiarize yourself with the Official Grand Prix Pinewood Derby car specifications.

Let Your Child Build the Car

It is impossible for your child to win the Pinewood Derby if you, the parent, build the car. Even if your car comes in first place, your child will know that he is accepting a trophy that you earned. The Pinewood Derby is about more than just winning; it's a time for children to learn skills and to develop a lasting relationship with an adult role model. Don't rob them of these opportunities. It is absolutely possible to build a winning car and still allow your child to do most of the work. Here are a few suggestions that I use every year as I build cars with my sons.

Build Two Cars. Building two cars will allow you to teach techniques by example and to make any mistakes on "your" car.

Build two cars

That's right—build two cars. When we build Derby cars, I build a car alongside my son. The technique I use is simple. My son watches me apply each step of the process to my car. Then, he does the same thing to his car. For example, I will demonstrate how to trace the car design onto my raw block of wood. When I'm finished tracing, he will trace the design onto his block of wood, having seen me do the same thing moments before. There will, of course, be steps that need to be performed by an adult for safety reasons. In these situations, try to find some way that your child can safely be involved in supporting the work you are doing. For instance, when I need to make a cut using a drill press and a forstner bit, I might let my son install the bit in the drill press.

Building two cars is also a speed tip that has helped us build faster cars and has saved us a lot of grief in the past. Because building mistakes and errors into a Derby car will not make it go faster, having two cars will allow you to make mistakes and have learning opportunities on "your car." Then, you can apply the techniques correctly to "his car." This is especially true the first time you attempt to apply a particular speed tip or method. After all these years of racing, we still build two cars every time.

Let your child help design the car

Your child will feel more connected to his car if he gets to help design it. Later in this book, I will discuss the issue of car design, which does have a significant impact upon speed. However, within the guidelines I lay out, there is room for individuality and creativity. Let your child be creative and influential in the design of the car.

Let your child pick the color and detail options

Once again, this part is a great way to keep your child involved. Let him pick the color of his car. If he decides to use stickers or decals, let him do the choosing. Color and stickers will not affect the speed of the car, so let your child have as much say in this part of the car design as possible. I remember one year that our car was quite interesting, to say the least. My son simply had to put non-matching stickers all over his car, and I let him do exactly what he wanted. The car looked kind of messy, but he liked it.

Teach principles while you work

Since the Pinewood Derby is, at least in part, devoted to developing woodworking skills, take the opportunity to teach your child as you work on this project together. For example, don't use precut shapes; instead, start with the standard block so that you and your child have the opportunity to design and cut the car together. Make sure he understands why each step of the process is important. If you ask my sons why we polish our car axles, they can tell you. The more your child understands, the more he will feel like he is a part of the process.

I always help my boys draw a pattern on a sheet of paper. They cut out that pattern and then transfer the design to the pine block. We then use either a coping saw or a band saw to cut the design into the block. By doing this, my sons are learning how to create and transfer a pattern. They also learn how to properly and safely use a saw to cut wood. I oversee this entire process and ensure that proper safety guidelines are adhered to. It's so important that your child be involved in the work and that he learns skills as he builds his car.

Teach the Processes. Make sure that your child understands each step in the process and what is being accomplished. The more he knows, the more he will feel as though he is part of the process.

The Three Basic Principles of Speed

There are three basic principles that govern how fast a car will go. Every speed tip in this book is designed to give your car the maximum benefit from each of these three principles.

Principle 1: Maximize potential energy

Gravity provides the energy that makes your car roll down the track and is a strong force that pulls on your car. You can think of gravity as the engine in your car. Before the race begins, your car is sitting stationary at the top of the track. It isn't moving yet. At this time, your car has what is called "potential energy." Potential energy is the amount of energy available to make your car roll down the track. The important point to understand here is that cars with more potential energy will roll faster as they move down the track, just as an automobile with a 300-horsepower engine will go faster than one with a 100-horsepower engine. The first principle of maximizing speed is to maximize the amount of potential energy your car has.

Believe it or not, 99% of the cars out there end up having far less potential energy than is actually possible. As you work through this book, I will show you how to maximize the potential energy of your car when the race begins. You will be the guy who shows up with the 300-horsepower engine.

Principle 2: Reduce friction

More Friction = Less Speed

Friction is the enemy of speed. You've built a car with tons of potential energy, but unfortunately not all of that energy actually gets converted into speed. Some energy is converted into heat as surfaces rub against one another causing friction. Most of the speed tips in this book are aimed at reducing friction. We want to save as much of your energy as possible so that it can be used to make your car rocket down the track.

Principle 3: Reduce wheel inertia

We want to reduce the inertia coefficient of the wheels as much as possible. What is inertia anyway? According to Newton's first law of motion, any stationary object will tend to remain stationary until it is acted upon by an outside force. In layman's terms, this law means that your wheels will stay motionless until a force pushes on them and makes them start moving. Objects at rest tend to resist a change in movement. This resistance to change is called inertia.

Heavy objects have more inertia than light objects. This means that lighter objects take a smaller force to make them start moving. For example, if you put lightweight wheels on your car, it will take a smaller force to make them start rolling down the track than it would if you put heavy wheels on your car. Cars with lightweight wheels will begin rolling sooner than cars with heavy wheels. This principle is discussed in more detail in Chapter 4, "Wheel Preparation," on page 61.

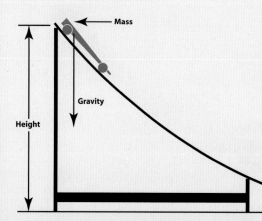

Potential Energy. Potential energy = mass × gravitational pull × height. Since we can't change gravity, we need to maximize height and mass—concepts that we'll discuss throughout this book.

The number of possible shapes and designs that can be applied to Pinewood Derby cars is endless. I'm sure you have some ideas floating around in your mind right now. However, before you begin, it is important to consider the following car design guidelines.

■ The dimensions of your car must remain within the width, height, length, and weight specifications listed in your Derby rules. The official Boy Scouts of America (BSA®) specifications are outlined on page 4 of this book.

■ Do not select a design with a pointed nose. A pointed nose will make it difficult for your car to rest on the pin at the starting gate. It may also cause your car to get bumped around when the pin drops, and it can create problems for electronic timing systems. In most cases, electronic timers require something that is at least 0.25 inches in width before the system will detect it. Some cars with a very pointed nose will not trip the timer until and inch or so of the car has passed under the sensor.

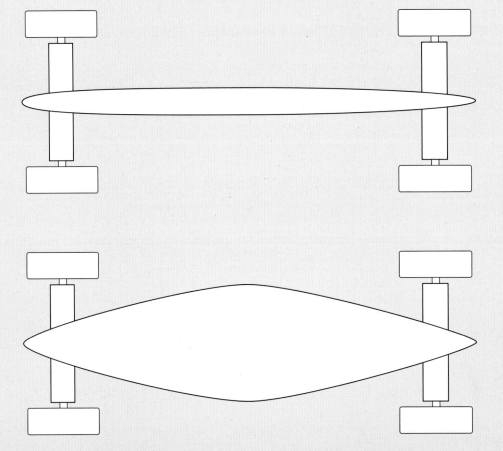

Avoid a Pointed-Nose Design. Pointed-nose designs, such as the ones shown here, can cause problems with the pins at the starting gate and with electronic timing systems.

■ Leave enough wood in the rear of the car so that you can place additional weight there. We will discuss weight placement in detail as we build the cars. However, it is important to know at the design stage that you will be putting most of your weight in the rear of the car.

■ Be sure that it is very clear which end of your car is the front and which end is the back. In many races, the race officials actually place each car on the track. I've seen many situations where the officials put the car on the track backward because they can't tell which end is which.

■ Choose an aerodynamic design. In other words, choose a design that allows the air to move over and around the car body in a smooth manner. Cars with aerodynamic profiles go faster.

■ Remember to let your child help design the car. This isn't "your" car. It's his.

An Aerodynamic Profile. An aerodynamic design allows the air to move over and around the car body in a smooth manner. The most basic aerodynamic design is the simple wedge (bottom). If you don't have a lot of time available, this is a time-proven design used by many Pinewood Derby winners.

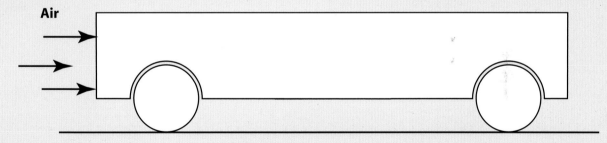

Air

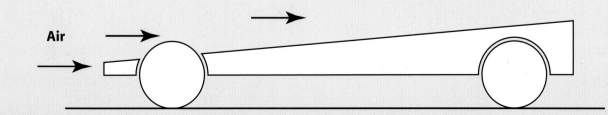

Air

Chapter 2
Building the Car Body

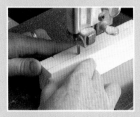   

The very first thing that you'll want to do is work on the design for your car. However, this involves more than just picking a good-looking design. Building the body of your car means laying the groundwork for the axle and wheel modifications that we'll be performing later on. This is also the time to concentrate on getting as much weight as possible in the rear of the car. You'll notice that I'm offering three different designs in this chapter—the Winning, the Champion, and the Ultimate—to match your skill level, whether you are competing in your first Pinewood Derby or are an experienced racer.

The Winning Car

This car is designed especially for people who are new to Pinewood Derby racing. Some beginners will open the box, pour out the contents, stare at the pile, and then look at each other, wondering what to do next. If that's how you feel, don't worry! I'm going to lead you through the process, step by step.

In order to build a competitive car, you must focus on four important speed components: weight placement, axle preparation, wheel preparation, and wheel lubrication. If you don't have very much time to devote to the building of your car, follow the Winning Car steps throughout the book (pages marked with the red bars and tabs) to concentrate on these basic principles of a competitive car.

As we build the car body in this section, we'll concentrate on cutting a simple but aerodynamic design, work with weight placement to maximize potential energy, sand and paint the car to help reduce friction, and lubricate the wheel well to further reduce friction.

Remember to follow the steps listed here in their numbered order. Because each level builds on the one before it, you may need to refer back to the previous sections to complete all of the steps for this level.

By sticking to the basics, you'll have a car that will hold its own, win some races, and make your child proud. Now, let's get started!

Materials and Tools

- Pencils
- Ruler
- Official Grand Prix Pinewood Derby Kit
- Band saw, scroll saw, or coping saw
- Vise
- Drill or drill press
- 25/64" bit (to fit the lead wire used in the demonstration)
- Sandpaper, 120 to 600 grit, several sheets of each
- Small digital scale
- Weights of choice
- Wood filler or wood putty
- Soft cloths
- Painter's tape
- Cardboard box, for homemade spray booth
- Spray-on primer, one can
- Steel wool, 0000 grade
- Spray paint of choice, one can
- Clear coat spray, one can
- Wheel lubricant, such as graphite or NyOil II®
- Safety goggles
- Dust mask

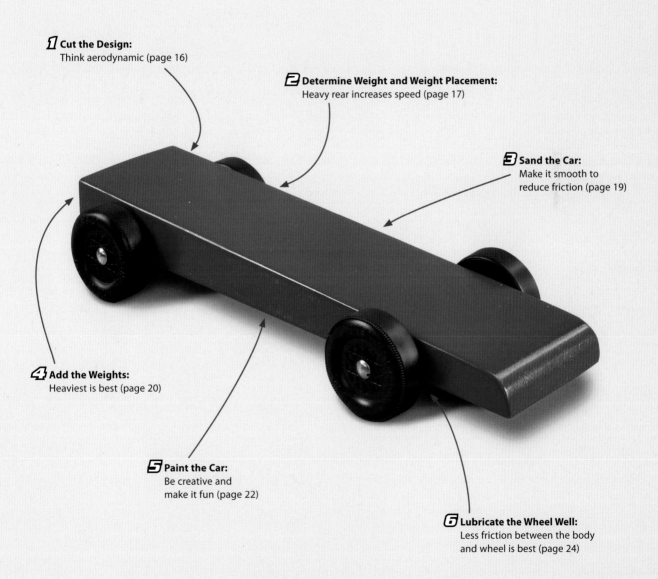

1 **Cut the Design:**
Think aerodynamic (page 16)

2 **Determine Weight and Weight Placement:**
Heavy rear increases speed (page 17)

3 **Sand the Car:**
Make it smooth to
reduce friction (page 19)

4 **Add the Weights:**
Heaviest is best (page 20)

5 **Paint the Car:**
Be creative and
make it fun (page 22)

6 **Lubricate the Wheel Well:**
Less friction between the body
and wheel is best (page 24)

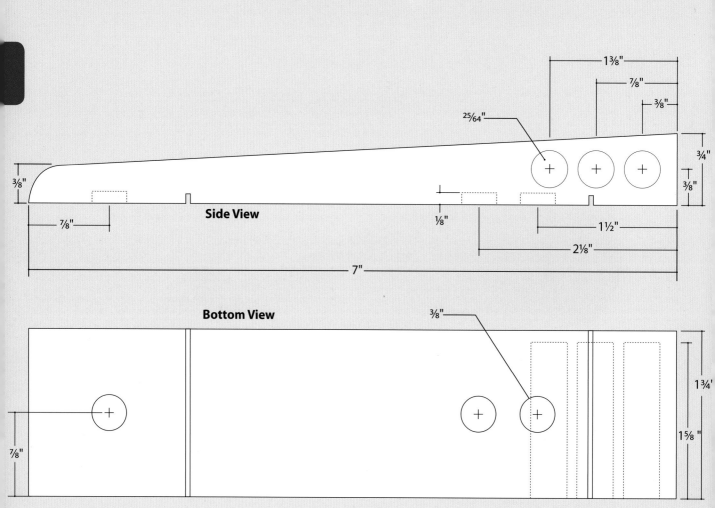

Side View

Bottom View

Cutting Guides

For this car, we will be using the basic wedge as our design. Since cutting this design basically involves one straight cut, we can simply measure the block, draw guidelines, and cut out the blank rather than use a template. However, if you'd like to use a template, one has been provided for you on page 15. If you choose a different design, you might find it useful to make a template before cutting the block.

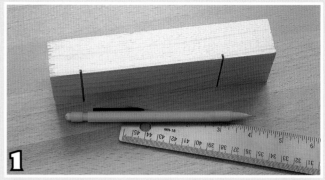

1

To prepare your wood for cutting, you'll need a pencil, a ruler, and the Pinewood Derby blank.

2

Make a mark ⁵⁄₁₆" up from the bottom of the car in the back and ⅞" up from the bottom of the car in the front (see the diagram on page 14 for more details).

3

Using the ruler, connect the two marks that you made. Now you are ready to cut the block.

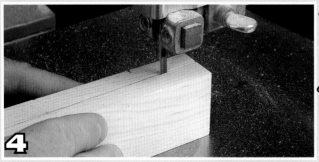

4

Cutting out the blank is easy with a band saw. Be sure to cut to the outside of your lines to allow for the kerf, or width, of the blade. Obey all safety rules and exercise caution when cutting the block.

Don't Have a Band Saw?

If you don't have access to a band saw, a coping saw can be used to cut the blank. Be sure to clamp the block securely in a vise. Cut the block halfway through; then, turn the block over, reclamp it, and saw the rest of the block.

The Winning Car

Determine Weight and Weight Placement

Correctly adding weight to your car is one of the most important things you can do to increase speed. The two most important rules dealing with weight and weight placement are 1) the heavier your car is, the faster it will go, and 2) rear-weighted cars go faster. By placing the weight in the rear of the car, you increase the potential energy of the car when it's at the starting gate. In other words, you have a larger gravity engine propelling your car down the track.

To place the weight in your car, drill holes for the weights toward the back of the car. In this demonstration, I will be drilling holes in the side of the car, close to one of the rear wheels. If you prefer, you can choose other placements for your weights (see the sidebar on page 18).

You will also need to create a way in which the weight can be adjusted. To do this, we will drill holes in the bottom of the car in the spots marked on the diagram. These holes will be used to fine-tune the weight after the car is painted and assembled.

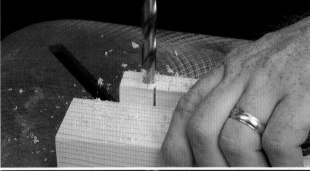

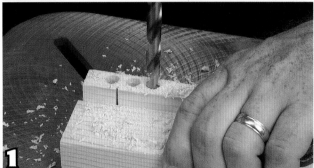

1 Draw guidelines for your weight holes on your blank. If you have a drill press, use it to drill the holes 1⅜" deep. Be careful not to drill all of the way through the block. (This demonstration is continued on the next page.)

Observing the Effect of Rear-Weighted Cars

The effect of placing the weight in the rear of the car is actually observable. Rear-weighted cars will appear to accelerate at the bottom of the track where the ramp levels out. I have actually had parents standing next to me lean over and say, "What's up with your car? It accelerates at the bottom of the track."

What really happens: A front-weighted car stops accelerating as soon as the front of the car reaches the flat portion of the track. It's no longer "falling." From that point on, the car is slowing down. A rear-weighted car will continue to accelerate until the back of the car reaches the flat portion of the track because the car "falls" until the entire car is located on the flat portion of the track.

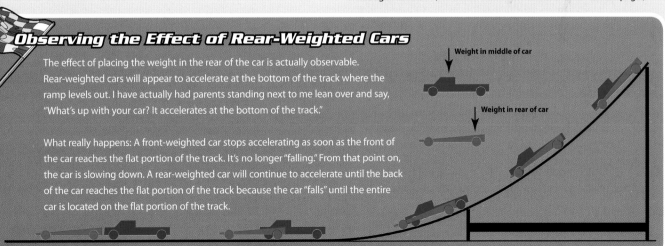

Weight in middle of car

Weight in rear of car

Determine Weight and Weight Placement

2 Then, mark the holes for weight adjustment as noted on the diagram. Drill them to a depth of 3/16".

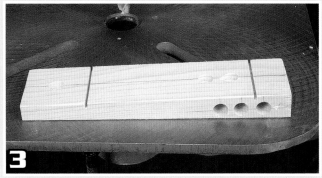

3 Your blank should now look like this.

Ways to Add Weight

If you'd like to choose a different location or a different type of weight other than the one shown in the step-by-step section, use these tips to position the weight properly. You'll need to consider the weight type and placement and the car's balance point to ensure that the weight is neither too far forward nor too far back.

Remember these two key rules:

- Put as much of your weight in the rear of the car as possible
- If you put the weight too far back, your car will pop a wheelie as it goes down the track, or it might even jump off the track

There are many ways to add weight to your car, so feel free to be creative in how you accomplish this part of the process. Just be sure to do it!

Here are just a few ideas:

- If you don't have access to a drill, glue the weights to the top of the car.
- If you decide to add a toy driver to the car, put him near the back to add weight in the correct location.

Once you have decided on the type of weight and its placement, find the balance point to make sure that your weight is located properly. The balance point, or the center of gravity, is simply the spot on your car upon which it will balance. You can easily locate the center of gravity for your car by placing it on a thin piece of wood, a ruler, or a wooden dowel. Slowly move your car back and forth until you find the right place. You'll need to continue checking the center of gravity throughout the process of adding weights to ensure that the balance point remains in the correct place.

Tips for balancing the weight:

- If the track you are racing on is a very smooth track, place your weight so that the center of gravity is 3/4" in front of the rear axle.
- If the track has bumps and rough spots, move the center of gravity forward so that it is located approximately 1" in front of the rear axle. This placement will give your car a little more stability when it hits those bumps.

Center of Gravity

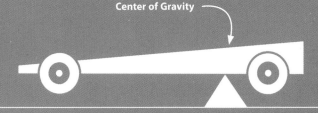

Sand the Car

Now that all of the drilling and cutting is complete, it's time to finish the basic shape of our car. This a great place for your child to participate in the process. We want to give the car a little bit of shape and sand it smooth. Remember, the car should move through the air as smoothly as possible, so the extra time that you put into this step is well worth the effort and could shave fractions of a second off your time.

1 Sand your car with several grades of sandpaper. I usually finish the process by using sandpaper with at least 220 grit to give it a smooth finish.

2 Take some time to give your design a little bit of style. Here I knocked off some of the sharp corners and rounded the nose of the car.

Add the Weights

Now that your car has its final shape, you'll need to add weights. In order to maximize potential energy, the car should weigh almost five full ounces (we'll add the final few tenths of an ounce at the very end), and the bulk of that weight should be located in the rear of the car. Any weight you place toward the front of the car is speed lost. Therefore, don't use weights that distribute the weight across the full length of your car.

Start by weighing your blank, wheels, axles, and weights of choice. If you don't have a small digital scale, borrow one or go to the local post office and ask them to weigh your car. Be sure that your car weighs a little less than 5 ounces. I like to aim for 4.8 ounces.

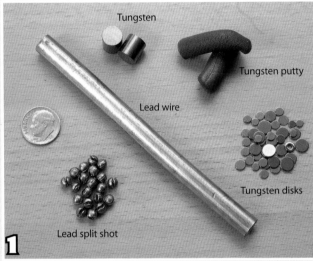

Insert the weights into the holes in your car.

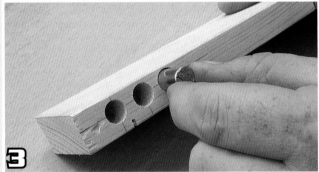

Tungsten

Tungsten putty

Lead wire

Tungsten disks

Lead split shot

Prepare your weights. Some types of weights require more work than others. Shown here clockwise from the bottom left is lead split shot, lead wire, tungsten, tungsten putty, and tungsten disks. The dime is shown for size reference. Since I have chosen lead wire, I will cut it to fit into the holes I drilled (see the Choosing Weights sidebar below for more information).

Choosing Weights

What type of weight should you use? Let's take a look at some of the different types.

Lead: For optimal speed, lead is one of the best choices. You simply cannot get enough weight in the rear of your car with almost anything else.

Zinc: Zinc weights are only a little more than half as dense as lead. If you use zinc weights, you will end up putting some of your weight toward the front of the car because the amount you'll use is so great. Weight in the front of the car will reduce speed.

Tungsten: Tungsten is much denser than lead and does not have lead's toxic properties. Though it can be very expensive, tungsten is also ideal for adding weight to very thin cars.

4 Prepare wood filler or wood putty, such as the Bondo shown here, for filling the holes. Use the filler or putty carefully—if you have too much weight at the official weigh-in, you'll have to find a way to remove some.

5 Apply the putty to the car, allowing it to fill the weight holes. You'll also notice in subsequent steps that I was able to fill some of the holes in the wood at the end of the block. Set it aside to dry according to the manufacturer's directions.

6 Once the putty has dried, sand it flush with the side of the car.

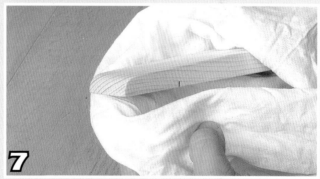

7 Give your car a final sanding with some 400-grit sandpaper to make it as smooth as possible. Wipe off any dust with a soft cloth to remove any remaining dust or particles that could become visible when you apply paint.

A Word about Handling Lead

Please remember that lead is a toxic substance. Wear gloves when handling lead, or be sure to wash your hands after you are finished handling it. Keep all lead products out of the reach of children. Follow any and all warnings that accompany lead products. Remember, do not use melted lead! Melting lead is dangerous, and there is no reason to use that method. You can safely add weight to your car using other lead products.

You have designed, cut, and sanded your car. It's starting to look cool! Now it's time to apply paint and perhaps stickers or decals. Remember that this is great place for your child to express his creativity.

I like to use spray paint because it goes on smoother than conventional paint. Also, a layer of spray paint is thinner than a layer of brush-on paint, and thus weighs less. Less paint means you have more control over where you put the weight of the car.

Cut some painter's tape to fit over the axle holes on your car. We don't want paint or anything else clogging the holes and adding friction.

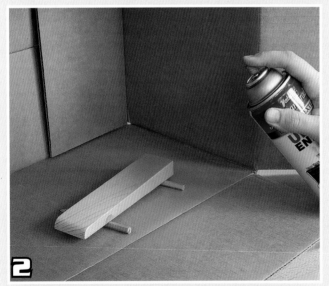

Set up a paint area using a cardboard box and two pencils. Place the car body on top of the pencils. Because the wood is full of small cracks and pores, it is a good idea to apply a coat of primer to fill the imperfections so the paint goes on evenly. Spray one good coat of primer on your car and let it dry. Be sure to follow the manufacturer's directions carefully.

Lightly sand the primer coat with 0000 grade steel wool. This will remove any bumps or particles in the primer coat. Wipe the car down with a cotton cloth.

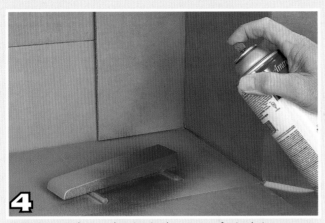

4 Now we are ready to apply paint. Apply one coat of paint, being careful not to apply it so thick that it runs. Let the paint dry. Follow the manufacturer's directions carefully.

5 Lightly sand with 0000 steel wool and wipe down with a cotton cloth.

6 Apply a second and a third coat of paint and let it dry. Then, apply one good coat of clear coat to make it shine!

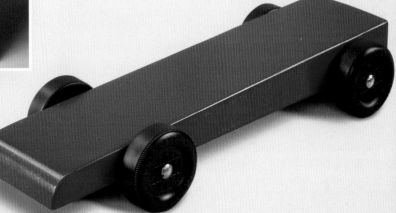

Remember, friction is the enemy of speed. One of the highest friction spots on your car is the small area where the wheel hub rubs against the car body. Every time contact is made in this area, you lose speed. So, you need to make this area as smooth as possible. We'll sand the area around the axle hole until it is extremely smooth and then add graphite to reduce friction.

Using Graphite

When using graphite, it is important to wear a dust mask. Because graphite particles are very light, they can become suspended in the air, where they get inhaled. Graphite is also one of those substances that tends to get all over everything. To contain the graphite, find a suitable place outside or cover your working surface with newspaper, cloth, or plastic.

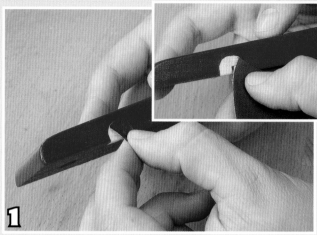

1 Once the paint is dry, remove the painter's tape and lightly sand the wheel well with 600-grit sandpaper.

2 Use painter's tape to cover the area around the wheel well to keep graphite from getting on the paint. Apply the graphite to a smooth piece of cloth.

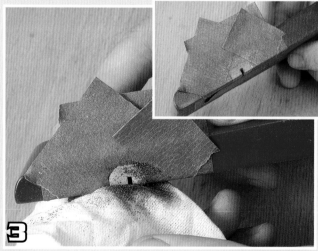

3 Using the cloth, gently rub graphite into the wood around the axle hole area. Continue to apply graphite until the area looks as smooth as it can be.

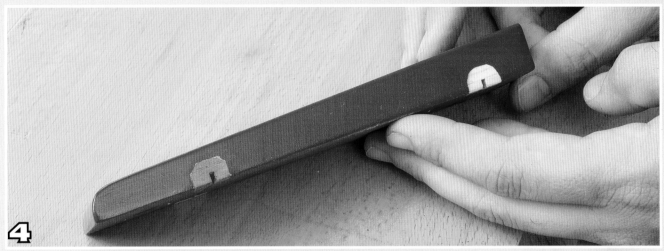

4

Check your work using a magnifying glass to be sure you don't have any big scratches or bumps. If you do, sand them down and reapply the graphite. Once you are happy with the area, remove the tape.

Removing Graphite from Paint

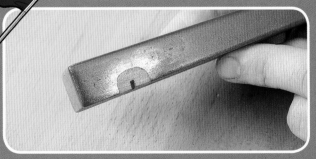

Though great for lubricating around the axle holes, graphite can get very messy if it comes in contact with the paint.

If you accidentally get graphite on the paint, use a little WD-40® on a rag to wipe off the extra. Do not spray the WD-40 right on the car because it could ruin your paint job.

The Champion Car

To build a champion, you need to do several things above and beyond what the rest of the people in your Derby have done. In this section, we'll take the car body to the next level by building on the steps that were performed for the Winning Car.

Remember, there is no magic step or secret that will turn your car into a winner. You win races by doing a whole list of things to your car, each of which might shave 1/1000 of a second off your race time. A lot of people will tell you to focus on the big speed tips and forget about the small ones, but performing the small steps in this section will save you fractions of a second—and they do add up.

In addition to performing the steps for the Winning Car—cutting a simple but aerodynamic design, working with weight placement, sanding and painting the car, and lubricating the wheel well—we will extend the wheelbase to gain even more potential energy and drill glue holes to ensure perfectly glued axles.

Remember to follow the steps listed on page 27 in their numbered order. Each level builds on the one before it, so you may need to refer back to the previous sections to complete all of the steps for this level.

OK, let's learn how to build a champion!

Materials and Tools

- Pencils
- Ruler
- Official Grand Prix Pinewood Derby Kit
- Band saw, scroll saw, or coping saw
- Drill or drill press
- Vise
- $^{25}/_{64}$" (to fit the lead wire used in the demonstration), #44, and $^{5}/_{64}$" bits
- Sandpaper, 120 to 600 grit, several sheets of each

- Small digital scale
- Weights of choice
- Wood filler or wood putty
- Soft cloths
- Painter's tape
- Cardboard box, for homemade spray booth
- Spray-on primer, one can
- Steel wool, 0000 grade
- Spray paint of choice, one can
- Clear coat, one can
- Graphite

- Clamp
- Square, carpenter's
- Piece of scrap wood, approximately 1¼" x 7" x 1¾"
- Stickers or decals of choice
- Scissors
- Sponge
- Small dish of water
- Safety goggles
- Dust mask

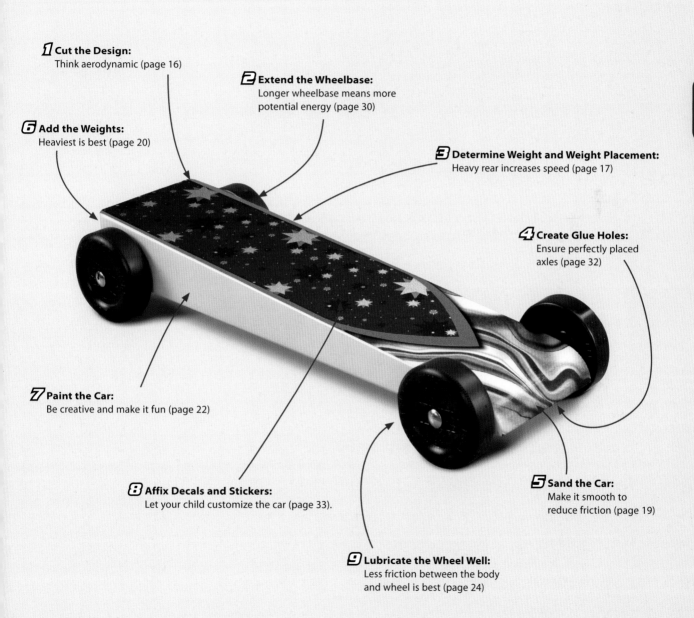

1 Cut the Design:
Think aerodynamic (page 16)

2 Extend the Wheelbase:
Longer wheelbase means more
potential energy (page 30)

6 Add the Weights:
Heaviest is best (page 20)

3 Determine Weight and Weight Placement:
Heavy rear increases speed (page 17)

4 Create Glue Holes:
Ensure perfectly placed
axles (page 32)

7 Paint the Car:
Be creative and make it fun (page 22)

8 Affix Decals and Stickers:
Let your child customize the car (page 33).

5 Sand the Car:
Make it smooth to
reduce friction (page 19)

9 Lubricate the Wheel Well:
Less friction between the body
and wheel is best (page 24)

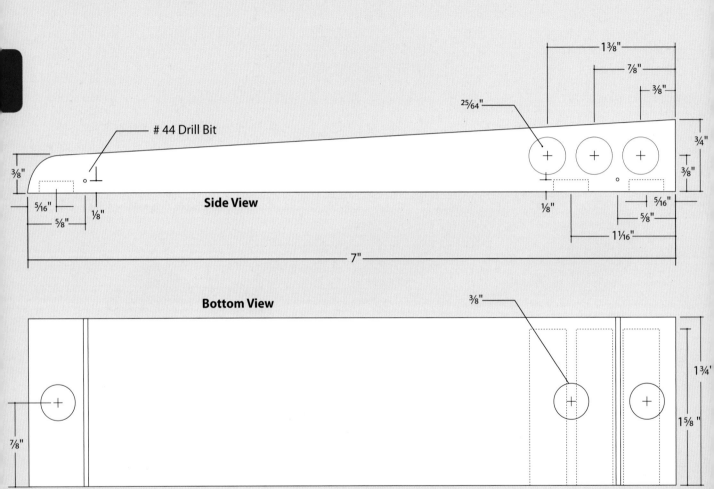

44 Drill Bit

25/64"

1 3/8"

7/8"

3/8"

3/4"

3/8"

3/8"

5/16"

5/8"

1/8"

5/16"

5/8"

1 1/16"

Side View

7"

Bottom View

3/8"

1 3/4'

1 5/8"

7/8"

Cutting Guides

Extend the Wheelbase

One of the most powerful things you can do to your car is to extend the wheelbase. That is, move the wheels so that the front wheels are closer to the front and the rear wheels are closer to the back. Remember to check your local rules to ensure the legality of this or any other speed tip before you start building your car.

This modification will give your car two powerful advantages over other cars. First, it allows you to place the weight more toward the rear of the car. The farther back you place the weight, the faster your car will go. If the weight is high on the track, your car gains potential energy. Remember: More Energy = More Speed.

Second, a longer wheelbase will make your car travel in a straighter line as it rolls down the track. The straighter it rolls, the less likely it will be to weave around and bump into the center guide rail. The shortest distance between the start and the finish is a straight line. Fewer Bumps = More Speed.

To accomplish this modification, you will need to drill a new set of axle holes in your wood block.

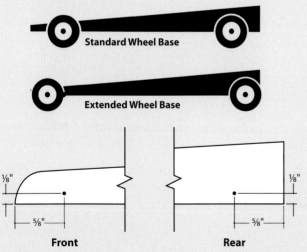

Standard Wheel Base

Extended Wheel Base

1/8"   1/8"

5/8"   5/8"

Front   **Rear**

The holes must be perfectly straight and aligned with each other on opposite sides of the car. This is best accomplished by using a drill press. You can also use the Derby Worx Pro Body Tool (see the sidebar on the next page). The holes need to be drilled ⅝" from either end of the block and ⅛" up from the bottom. Distances shown are measured to where the middle of each hole should be located.

1 If you are using a drill press, set up a jig to hold the block in place while you drill each hole. Proceeding without the jig almost guarantees that you will not be successful in drilling the holes correctly. The idea here is to create a repeatable process so you know exactly where the hole is going to be drilled when you put your car block on the drill press table. Building a jig can be as simple as clamping a square on the drill press table, as I have done here.

2 Once you have the jig in place, drill some practice holes in another block of wood. Measure the location of the hole and be sure that it is exactly where you want it to be. When you are confident in the location of the hole, you can proceed to drill the holes in your actual car block.

3 Drill each hole using a #44 bit. Do not try to use a long drill bit to drill one hole that goes all the way through the block. A #44 drill bit is thin and weak, and, in most cases, it will not drill a straight hole all the way through the block. Instead, the bit will wander or curve slightly. When this happens, the drill exits the opposite side of the block slightly out of position.

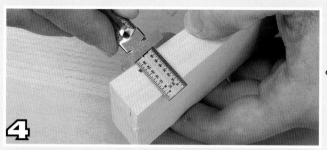

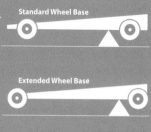

Center of Gravity

Please note that when you extend the wheelbase, the center of gravity rule still applies. That is, the center of gravity for your car should be between ¾" and 1" in front of the rear axle. When the wheelbase is extended, that location will be located farther to the rear of the car.

Standard Wheel Base

Extended Wheel Base

4

When all four holes are drilled, measure them all one more time to be sure they are correctly placed. Measure up from the bottom of the car as well as from the front or back of the car.

Using the Pro Body Tool to Drill Axle Holes

If you don't have drill press, you might consider using the Derby Worx Pro Body Tool to drill perfectly aligned axle holes using your handheld power drill. The required #44 drill bit is included with the tool.

1

Test fit the tool on the bottom of the block with the two "ears" positioned on the sides. If the tool will not fit onto the block, use coarse grit sandpaper to reduce the width of the block until the tool fits snugly. If the tool is loose, tighten the fit by placing a piece of paper—folded as needed—between one "ear" of the tool and the block.

2

Measure and mark the position of the new axle holes. You can use a square to draw a straight line starting at the middle of the axle slot and extending to the top of the car. Or, measure the distance from one end of the block to the middle of an axle slot and transfer that measurement to the top of the block. Use a ruler to draw a line from that measurement to the center of the slot.

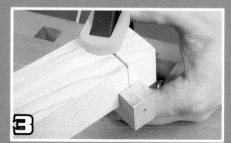

3

Position the tool on the bottom of the car with the two ears on the sides of the block. Align the index mark on the tool with the line corresponding to the rear axle slot or hole on the block. Clamp the tool in place.

4

Hold the block firmly, position the drill bit in the hole, and slowly drill ¾" into the block. Without releasing the clamp, turn the tool and block over, and drill the second hole.

5

Repeat these steps for the other holes in the block.

Create Glue Holes

If you choose to extend the wheelbase of your car, I highly encourage you to pay attention to this little tip. One of the worst things that can happen on race day is to have a wheel get knocked off the car. Truthfully, I've seen this happen a lot of times. Similarly, if you spend time aligning the wheels, then you need to be absolutely sure the axles will stay put once you have them aligned properly. You don't want to have your car get bumped a little in the first race and then be totally out of alignment. Hence, the axles need to be firmly glued in the holes. A lot of people end up killing their prospects of winning when they try to glue the axles. If any glue whatsoever gets on the wheel hub or on the car body where the hub will touch, you will lose significant speed. The glue will act like sandpaper and create friction. Here's a technique that we have used for years to get perfectly glued axles.

Simply drill two small 5⁄64" holes through the bottom of the car body and into each axle hole.

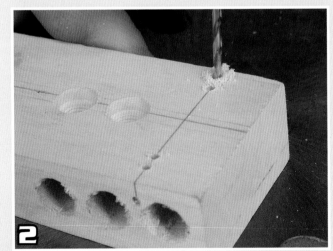

When it's time to insert the axles, you will be able to apply one drop of cyanoacrylate, or CA, glue to each glue hole. Allow the glue to dry overnight before turning your car right side up. The glue will cement the axles firmly without the risk of getting glue anywhere near the wheels.

Affix Decals and Stickers

Many children want to add decals or stickers to their cars. It really doesn't make any difference in the long run which decals or stickers you apply (if any). To give you an idea of what can be done, I'll be using a full-car decal for this demonstration. It's a large decal that you transfer by applying water. The car looks like it has had a professional paint job. Check your local Scout Shop or *www.scoutstuff.org* for this type of product.

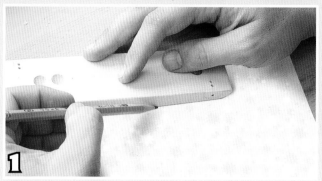

Once you have painted your car, choose your decal and trace the shape of your car onto the back of the decal.

Using scissors, cut out the decal, and remove the backing.

Fold the decal over the car.

Use a sponge to rub a small amount of water on the decal.

Remove the other backing and rub out any wrinkles with your fingers.

The Ultimate Car

If you are ready to go the extra mile and take home the first-place trophy, this section is for you. Here, we'll lay the groundwork for a car that leaves all the others in the dust. Contained in this section are all of the tips you'll need to complete to make your car as fast as it can be.

To make the Ultimate Car, we'll go beyond what was done to the Champion Car. In addition to working with weight placement, sanding and painting the car, lubricating the wheel well, extending the wheelbase, drilling glue holes, we'll bake the block to get as much weight as possible in the rear of the car, cut an advanced design to further maximize weight placement, prepare to have only three wheels on the track to reduce friction, and create a "quick start" design to gain speed.

Remember to follow the steps listed on page 35 in their numbered order. Each level builds on the one before it, so you may need to refer back to the previous sections to complete all of the steps for this level.

Let's get started!

Materials and Tools

- Pencils
- Ruler
- Official Grand Prix Pinewood Derby Kit
- Band saw, scroll saw, or coping saw
- Drill or drill press
- Vise
- $^{25}/_{64}$" (to fit the lead wire used in the demonstration), #44, and $^{5}/_{64}$" bits
- Sandpaper, 120 to 600 grit, several sheets of each
- Small digital scale

- Weights of choice
- Wood filler or wood putty
- Soft cloths
- Painter's tape
- Cardboard box, for homemade spray booth
- Spray-on primer, one can
- Steel wool, 0000 grade
- Spray paint of choice, one can
- Clear coat, one can
- Graphite
- Clamp
- Square, carpenter's

- Piece of scrap wood, approximately 1¼" x 7" x 1¾"
- Stickers or decals of choice
- Scissors
- Cookie sheet
- Clear tape
- Paper clip
- Cyanoacrylate, or CA, glue
- Small rectangle of electrical tape
- Safety goggles
- Dust mask

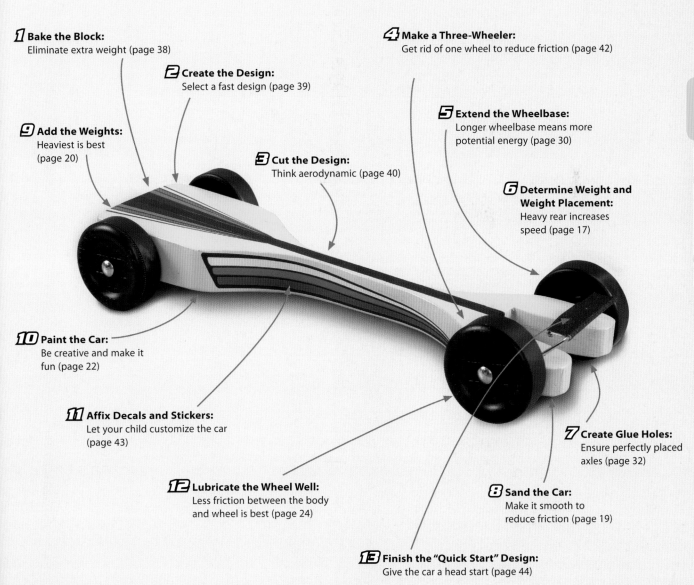

1 Bake the Block:
Eliminate extra weight (page 38)

2 Create the Design:
Select a fast design (page 39)

9 Add the Weights:
Heaviest is best
(page 20)

3 Cut the Design:
Think aerodynamic (page 40)

4 Make a Three-Wheeler:
Get rid of one wheel to reduce friction (page 42)

5 Extend the Wheelbase:
Longer wheelbase means more
potential energy (page 30)

**6 Determine Weight and
Weight Placement:**
Heavy rear increases
speed (page 17)

10 Paint the Car:
Be creative and make it
fun (page 22)

11 Affix Decals and Stickers:
Let your child customize the car
(page 43)

7 Create Glue Holes:
Ensure perfectly placed
axles (page 32)

12 Lubricate the Wheel Well:
Less friction between the body
and wheel is best (page 24)

8 Sand the Car:
Make it smooth to
reduce friction (page 19)

13 Finish the "Quick Start" Design:
Give the car a head start (page 44)

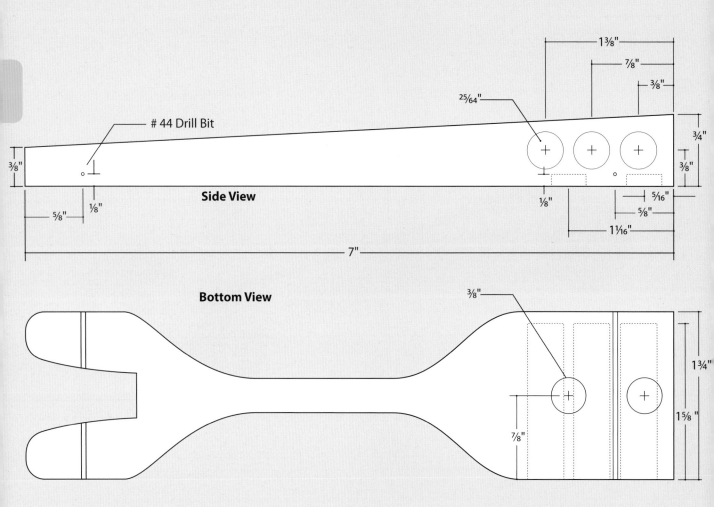

1³⁄₈"

⁷⁄₈"

³⁄₈"

²⁵⁄₆₄"

44 Drill Bit

³⁄₄"

³⁄₈"

Side View

³⁄₈"

¹⁄₈"

³⁄₈"

⁵⁄₁₆"

⁵⁄₈"

¹⁄₈"

⁵⁄₈"

1¹⁄₁₆"

7"

Bottom View

³⁄₈"

1³⁄₄"

1⁵⁄₈"

⁷⁄₈"

Cutting Guides

I have said several times that the more weight you put in the rear of the car, the faster the car will go. However, did you know that even the initial block contains weight that can be moved to the rear? All wood blocks contain water, and water is a very heavy substance. Just a little bit of it can weigh a lot, relatively speaking. All of the water locked up in the front half of your car is badly placed weight, so move it!

Before you do anything else to your car, put the wood block in your kitchen oven and bake it. This will cause most of the water locked up inside the wood to evaporate. With that water gone, you will be free to put more weight near the back of the car. Children should not perform this task without adult supervision.

Preheat the oven to 250 degrees. Depending on the humidity in your area, you may need to adjust the temperature or the time you use for baking the block. However, even in the Sahara Desert, a wood block will have some water in it. Every gram of weight you move to the rear of the car will make your car go faster. If you want to bake more than one block at a time, you may need to adjust the cooking time a little bit.

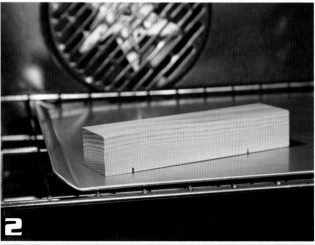

Once the oven is preheated, place the block on a cookie sheet and then right into the oven. We usually bake our block for about 2 hours.
Caution: Be sure to keep a close eye on what's going on in the oven. If you neglect the block, you could start a fire or overbake the wood. If the wood begins to turn brown or black, you have the temperature too high. We want to evaporate the water, not burn the wood.

Create the Design

When it comes to Derby racing, car design is a controversial topic, but it shouldn't be! There are car designs which produce faster cars than others, and that's a fact. If you want to build the Ultimate Car, then you need to select a fast design. So, what are the key factors of a fast car design?

Design Factor #1: Choose an aerodynamic design. We have discussed this principle on page 9. However, I wish to emphasize this point again. Use a design that has a low, aerodynamic profile. Aerodynamics do make a difference in speed.

Design Factor #2: Maximize the amount of weight placed in the rear of the car. This means removing as much wood from the middle and the front of the car as possible. Remember, any weight placed in the middle or at the front of the car is badly placed weight. Moving that weight to the rear of the car will increase speed!

This is one of those speed tips to which almost no one pays attention. Think back to last year or pay attention when you see all of the cars this year. How many people made any attempt whatsoever to remove wood from the middle or the front of the car? Every sliver of wood or pinch of sawdust that you move from the middle to the back of the car will help increase the speed.

Design Factor #3: Use what I call a "quick start" design (see diagram at right). How would you like it if your car had already started rolling down the track before any of the other cars even started moving? To the casual observer, it will simply look like your car is just really fast off the starting line. However, the truth is that your car really does start rolling down the track before the others even have the chance to move.

This is accomplished by cutting a channel in the nose of the car. A bar is then positioned above that channel. As soon as the starting pin drops below that bar, the car becomes free to start rolling.

We have raced in four Cub Scout packs in two states. Every single one of these packs has allowed us to use this speed tip. It's been a wonderfully powerful advantage! However, if your local pack will not allow you to implement this technique as described above, there is a simplified version that will still give you a quick start and should be legal anywhere.

Instead of cutting a channel in the nose, simply undercut the nose. As soon as the starting pin drops below the lip, the car will begin to roll.

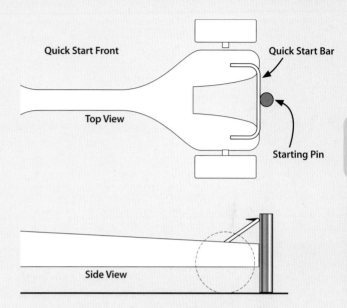

Quick Start Front — Top View — Quick Start Bar — Starting Pin

Side View

Alternate Quick Start Front

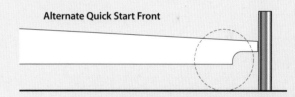

Alternate way to use the quick start design, shown from the bottom of the car.

The Ultimate Car

Once you have selected your design, you will need to make a template that can be used when cutting the block. Most designs will require that you make a template for the top view of the car and another template for the side view. If you don't want to create your own design, you can use the patterns that we've provided on page 37.

1

Cut out your design template and attach it to the block of wood with clear tape. Do not glue it on. Glue will just create a mess that needs to be sanded off later. I normally attach the side view first. However, whether you attach the side view or the top view first will depend upon which makes the most sense when cutting out the car.

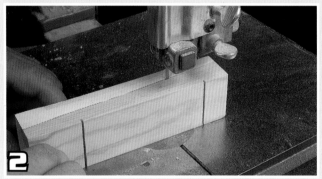

2

With the side template attached, you are ready to cut the block. I recommend using either a band saw, a scroll saw, or a handheld coping saw. Be sure to obey all safety rules and exercise caution when cutting the block. Find a way your child can safely help in this process.

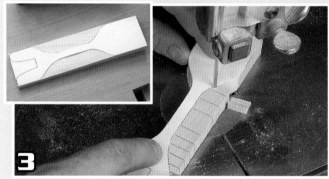

3

After cutting out the side view, attach the top view template. Then, cut out the rest of the car body. Notice that I am cutting straight in on the sides before I cut along the template line. This will give the band saw smaller chunks of wood to remove. If you don't have a band saw, this same technique will making cutting easier on the coping saw or the scroll saw.

4

Complete the process by sanding your car with several grades of sandpaper. Sand through at least 220 grit to give it a smooth finish.

Making the Template Easier to See

If you are having trouble seeing the template, try coloring it before you attach it to the block. It's important that you are able to clearly see where you are cutting when you are at the band saw.

Improving Standard Designs

While some designs inherently possess the features that make them faster, it is possible to take a standard design with standard weights and improve it by making some small changes. Because I've seen a lot of coupe designs over the past several years, we'll use that as an example.

By making the two modifications shown below, you will have a coupe design with the weight located in the rear of the car, which will dramatically improve its performance. These same techniques can be used to improve many of the standard designs you see at local Pinewood Derbies. So, if your child has his heart set on a design that isn't inherently fast, you can use these tips to maximize the design's potential.

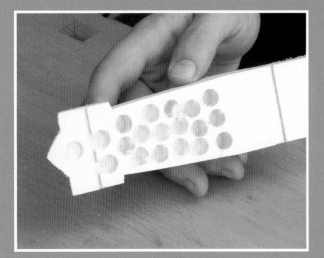

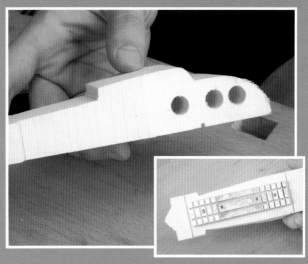

Drill out wood in the center of the body: Fast cars have as much wood removed from the center of the car as possible. This allows for more weight to be placed in the rear of the car. In this photo, I have used a drill to remove wood from the center of the coupe. All of this weight is now free to be relocated to the rear of the car.

Put the weight in the rear: Many people add weight to the car by attaching a long bar of metal to the bottom. Instead, drill three holes in the rear and fill them with lead. If you use a $\frac{25}{64}$" drill bit, you will be able to easily slide standard lead wire into the holes.

This is an awesome speed tip! We'll be marking our drill holes so that only three of the wheels actually ride on the track. The fourth is set about ⅟₁₆" higher so that it never actually touches the track itself. Remember our favorite Pinewood Derby rule? Friction is the enemy of speed. Well, two major sources of friction are the contact the wheels make with the track as the car rolls and the wheel axle contact we have discussed before. You can reduce that friction by placing only three wheels on the track to begin with. If the wheels on your car are aligned and balanced, then you only need three wheels. The wheel you elevate needs to be one of the front wheels.

If you are mixing and matching the tips in this book and you decide to use the standard wheelbase instead of the extended one, you can still use this tip. Simply cut the standard wheel groove a little deeper so that the axle can be inserted higher up on the block.

Drilling the Raised Hole with the Pro Body Tool

If you decide to use the Derby Worx Pro Body Tool to drill your axle holes (see the sidebar on page 31), drill the regular three axle holes first. Then, transfer the line on the side of the block to the bottom of the block. Position the tool on the desired side of the block, and align the index mark on the tool with the line on the bottom of the block. Make sure that the third hole on the tool is positioned near the bottom of the block. Clamp the tool in place and drill the raised hole.

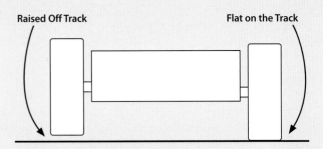

Raised Off Track **Flat on the Track**

Decide which front wheel you want to raise and mark that axle hole.

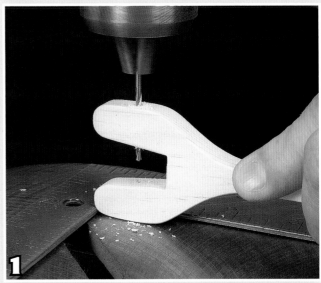

1

If you are using a drill press, it's a good idea to set up a jig to hold the block in place when you drill the hole. Plus, you can then adjust the jig to use for the other three holes, as shown in the next process on page 30. Building a jig can be as simple as clamping a square on the drill press table, as I have done here. Use a #44 bit to drill the hole. **Note:** If you choose to use this design or a similar design, you will need to trim the axles to fit the channel cut in the nose.

Affix Decals and Stickers

Many children want to add decals or stickers to their cars. It really doesn't make any difference in the long run which decals or stickers you apply (if any). For the Ultimate Car, I'll be using dry transfer decals to show you some of the possibilities.

Dry transfer decals look much more professional than any kind of sticker I've ever seen. We have used dry transfer decals for several years. Every year people come up and ask where we got "those awesome-looking decals." They actually end up looking like you painted the design on the car. Stickers look like . . . well, stickers. You can find dry transfer decals at any major hobby store, your local Scout Shop, or *www.scoutstuff.org*.

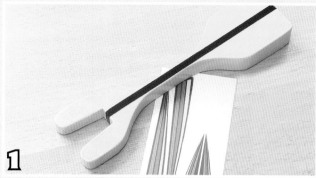

1 Once you have painted your car, choose the dry transfer decals that your child likes the most.

2 Using scissors, cut out the decal and position it on the car.

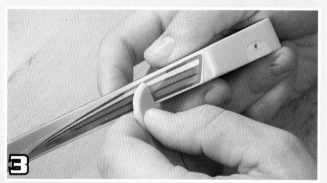

3 Hold the decal securely in place. Then, use any smooth object to burnish the decal onto the car.

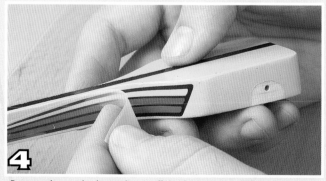

4 Remove the overlay by peeling it off slowly at a steep angle.

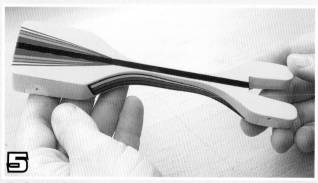

5 The finished design.

The Ultimate Car

If you have cut a channel in your car, create a bar over the channel that looks like a natural part of your car design. The bar also makes sure that the laser timer picks up your car's winning time! **Note:** We completed these steps with the wheels placed (not glued) for fit. You can complete this step with or without the wheels in place.

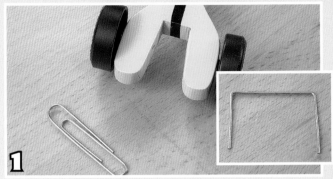

1

A simple paper clip makes a great bar. Start by bending it into the shape of a U.

2

Use the U-shaped paper clip to mark its position on the car.

3

Using a handheld drill with a bit just large enough to allow the wire to fit into it, drill two small holes where you made the marks. The holes should be about ¼" deep.

4

Glue the paper clip in place using CA glue or 60-second epoxy. Let it dry overnight.

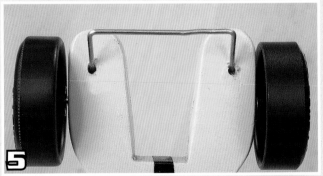

5

Be sure your bar doesn't extend beyond the front edges of your car, as this will change your car's dimensions and might get you in trouble with your local race officials.

6

A small rectangle, like this piece of electrical tape, will create a little more width for the bar, making it easier for the timer to see.

A Word about CA Glue

CA glue has many uses and can come in handy when building Pinewood Derby cars. However, be very careful when handling CA glue, the fumes of which can irritate the eyes and sinuses. Be sure to work in a well-ventilated area and do not bend too closely over your project. If you glue your fingers to each other or to objects in your workspace, acetone or nail polish remover will help remedy the situation.

Chapter 3
Axle Preparation

Correctly preparing your axles is the single most important thing you can do to increase the speed of your car. Remember, friction is the enemy of speed. The largest source of friction is the contact made between the axles and the wheels. As we did in the previous chapter, "Building the Car Body," we'll be going over three levels of axle preparation. The Winning axles will show you the basic steps you need to perform to make your axles competitive. The Champion axles will take your axles a step further and really increase your car's speed. The last section, the Ultimate axles, will maximize your speed potential.

The Winning Axles

In this section, I will introduce the two steps in axle preparation that you must complete to have a competitive car—removing burrs and polishing your axles. All Official Pinewood Derby axles have burrs, which must be sanded off to decrease friction. Polishing your axles will also cut down on friction. As the car moves down the track, the wheels turn and rub on the axle shaft continually. It is essential that the axle surface be as smooth and as shiny as possible. You need to polish, polish, polish those axles until they shine like mirrors! All competitive cars have polished axles, but some are polished better than others. The method I'm demonstrating here has been used for several years and will give your axles a better polish than most other methods.

Keep in mind that this is a great place for your child to do a major portion of the work—just be sure to teach principles as you work. It's important that he understands why each step in the building process is being performed. You might even have a contest to see who can make their axles shine the brightest! Remember to make the building process fun.

Materials and Tools

- **Electric drill**
- **Wet/dry sandpaper, 400, 600, 1000, 1200, 1500, 2000, 2500, and 3000 grit, several sheets of each**
- **Small dish of water**
- **Small triangular file**
- **Vise**
- **Magnifying glass**
- **Safety goggles**
- **Dust mask**

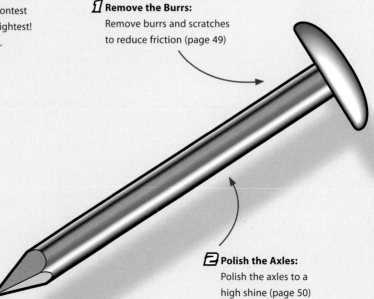

1 Remove the Burrs:
Remove burrs and scratches to reduce friction (page 49)

2 Polish the Axles:
Polish the axles to a high shine (page 50)

Remove the Burrs

All Official BSA axles, when they first come out of the box, have a set of small burrs located underneath the head of the axle. Whenever the wheel and the axle head touch, these little burrs create a large amount of friction. The burrs are almost like minibrakes, so they must be removed. The head of the axle must be polished smooth as well. Remember that friction is the enemy of speed, and these little burrs are a major source of friction.

1

Notice the burrs that are present on all official Scout axles, which come in the box with the rest of the materials. We will be removing them using an electric drill and a small triangular file.

2

Clamp your drill in a vise. Be careful not to tighten the vise too much; you might damage your drill by doing so.

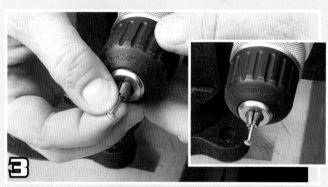

3

Insert an axle into the drill, leaving about half an inch exposed. Be sure that the axle is tightly secured in the drill.

4

Turn the drill on medium to high speed. Then, using a triangular file, remove the burrs. Put only light pressure on the file so you do not damage the axle in the process of removing the burrs.

5

The finished axles should look like the one shown here. Keep your axle in the drill for the next steps.

The Winning Axles

A great deal of friction is created when the wheel rubs against the axle while it is turning. To reduce this friction, make the surface of the axle as smooth as it can be. Believe it or not, the outcome of a race may well depend upon who does the best job polishing their axles.

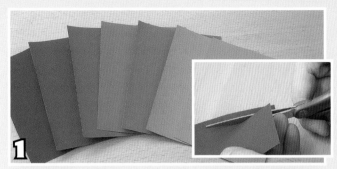

1 Have a number of different grits of wet/dry sandpaper on hand. Select a sheet of 400-grit sandpaper to start, and cut a small strip about ¼" wide and 4" long.

2 Turn the drill on medium to high speed. Dip the strip of sandpaper into a small dish of water, and then apply sandpaper to the axle. Be sure to sand the entire axle, including the inside surface of the head. This step should take about 15 seconds.

3 Turn off the drill. Look at the axle with a good magnifying glass. Are there any deep scratches left? If so, then turn the drill back on and polish some more! Do not move to the next step until all of the deep scratches have been removed.

4 Repeat Steps 2 and 3 using a strip of 600-grit sandpaper. Look at the axle with a magnifying glass after each step to be sure the axle shaft is as smooth as it can be.

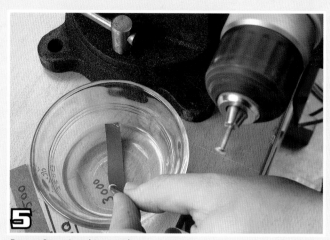

5 Repeat Steps 2 and 3 several more times using 1000-, 1200-, 1500-, 2000-, 2500-, and 3000-grit sandpaper.

6 When you've finished sanding, your axles should look like this.

Polishing with Pumice

Traditional axle polishing calls for using a paste made of pumice after the 600-grit sandpaper, but this is a bad idea—for a couple of reasons. In addition to being messier than using wet/dry sandpaper, moving from 600-grit sandpaper to pumice (equivalent to 1800-grit sandpaper) is too big a jump. The pumice is too fine to tackle the scratches left behind by the 600-grit sandpaper. So stay away from the pumice and stick with my method of using successively finer-grit sandpaper. Your axles will shine so brightly, you'll need sunglasses to look at them!

The Champion Axles

To create Champion axles, we will remove the burrs and polish the axles, just as we did for the Winning axles. But, don't stop there—you can do much more! Once we have removed the burrs, we can further reduce the friction between the axle and the wheel by tapering the axle head. The idea here is to reduce the surface area of the axle head so that only a small portion of the head can make contact with the wheel. Ideally, you will taper the axle head and remove the unwanted burrs at the same time.

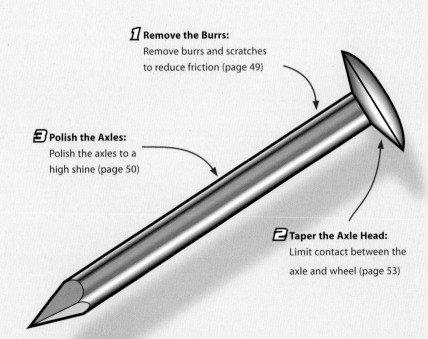

1 Remove the Burrs:
Remove burrs and scratches to reduce friction (page 49)

3 Polish the Axles:
Polish the axles to a high shine (page 50)

2 Taper the Axle Head:
Limit contact between the axle and wheel (page 53)

As the axle head and the wheel rub against each other, they create friction that slows your car down. You can minimize this friction by tapering the head of the axle with an electric drill and a small triangular file.

1 Using a smooth-tooth file, gently file the axle head to an angle of about 30 degrees.

2 Select a sheet of 400-grit sandpaper, and cut a small strip about ¼" wide and 4" long. Dip the strip of sandpaper into the small dish of water, and then apply it to the axle to polish the taper that you just filed. Then, continue with the Polish the Axles instructions found on page 50.

The Champion Axles

The Ultimate Axles

Even though the steps for the Ultimate axles include removing the burrs, tapering the head, and polishing the axles, this set of axles goes even further.

Set your car apart by starting with the straightest axles you can find. We can also decrease the friction between the axle and the wheel by limiting the surface area that's touching the wheel.

Remember to get your child involved in these steps. Making sure that your child understands why each step is important allows him to feel like he's a part of the process. Also remember to check your local rules and regulations before modifying axles.

Materials and Tools

- Electric drill
- Wet/dry sandpaper, 400, 600, 1000, 1200, 1500, 2000, 2500, and 3000 grit, several sheets of each
- Small dish of water
- Small triangular file
- Vise
- Magnifying glass
- Smooth-tooth file
- Drill press (optional)
- Flat file
- Safety goggles
- Dust mask

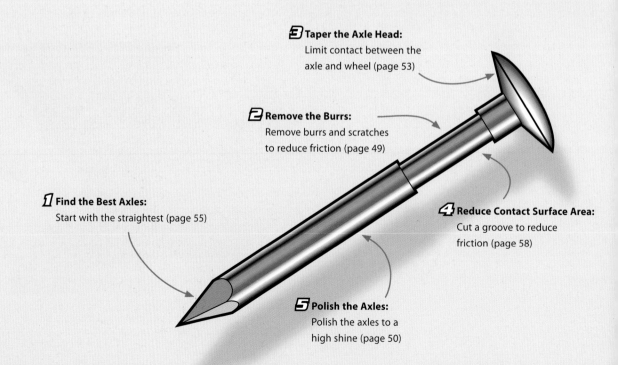

3 Taper the Axle Head:
Limit contact between the axle and wheel (page 53)

2 Remove the Burrs:
Remove burrs and scratches to reduce friction (page 49)

1 Find the Best Axles:
Start with the straightest (page 55)

4 Reduce Contact Surface Area:
Cut a groove to reduce friction (page 58)

5 Polish the Axles:
Polish the axles to a high shine (page 50)

Find the Best Axles

The title of this section pretty much says it all. Simply stated, there's no such thing as a straight axle that comes directly out of the box, but some are much straighter than others. This concept is very important to understand. If you use axles that are curved or bent, your wheels will wobble as they go down the track. Bent axles will also affect the alignment of your wheels. Simply put, Bent Axles = Less Speed.

The difference between a good axle and a bad axle is not normally visible to the naked eye. Using a drill press or a handheld drill, however, we can find the best axles of the bunch.

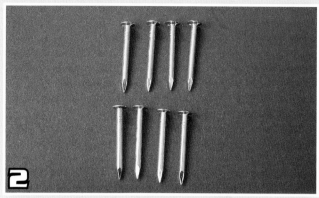

2

Here we're using eight as an example. Can you tell the difference between the axle sets? The axles on the bottom are not straight. Your car will have less speed if you use them. The axles on the top are actually quite close to being straight. They're about as good as you can get out of the box.

1

In order to find a good set of straight axles, you will need to start with about 20 raw axles. To get additional axles, you'll need to buy extra axle/wheel sets. Use only Official Pinewood Derby axles for pack races.

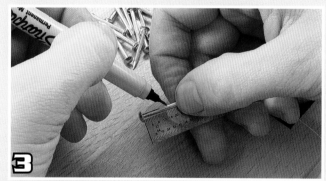

3

Take each axle and put a small mark on it about ½" down from the sharp, pointed end.

4

Clamp your power drill in a vise or use a drill press. (This demonstration is continued on the next page.)

The Ultimate Axles

5 Insert one axle into the drill or drill press, clamping the axle at the previously marked location. This will ensure that each axle is clamped at the same relative location during this process. Then, turn the drill on medium to high speed and observe the spinning axle. Bent axles, like the one shown here, will wobble as they turn in the drill.

6 Relatively straight axles, like the one shown here, will wobble only a little. Please note that all axles will wobble to some degree. The idea here is to pick the four axles that appear to wobble the least amount. These will be the four straightest. If you start with 20 axles, you will be able to find four that are really quite good.

Axle Hardening

The idea behind axle hardening is to heat an axle to "red hot" using a propane torch. You then drop the axle in water to cool it off quickly. This hardens the metal, making the axle scratch resistant.

On the surface, this technique sounds really cool. It's so "high tech." However, I've tested this method several times, and there is a negative side effect to this method, which nobody seems to have noticed. The hardening process can warp the axles. If you observe "hardened" axles spinning in a drill press, the effect is obvious. In my opinion, axle hardening does little to help you win the Derby and a lot to help you lose. Plus, it's dangerous!

Using the Derby Worx® Pro Axle Press

You can further straighten and improve your axles by using the Derby Worx Pro Axle Press. This nifty tool enables you to straighten bent axles. Simply put your axle in the press, tap it with a mallet, and—presto!—you have a straight axle. It works best if you start out with the best axles you can find, so be sure to follow the instructions in the Find the Best Axles section. I've used this tool for several years and find it to be everything it claims to be.

1

Insert an axle into the chuck of a drill. Use a file to remove the burr under the nail head and any crimp marks on the axle shaft.

2

Make a mark anywhere on the head of the axle.

3

Fully insert the axle, point first, into the axle press. Rotate the axle head so that the mark is located at the top (the 12 o'clock position), and close the press.

4

Place the press on a solid surface, hold it in place, and strike the top of the press four to six times with a hammer. Don't strike too hard; medium strikes are fine.

5

Open the press and repeat Steps 3 and 4 with the mark at the 4 o'clock position.

6

Open the press and repeat Steps 3 and 4 with the mark at the 8 o'clock position.

7

After the last strike at the 8 o'clock position, lay the press on its back (axle head upward) and strike the axle head two to four times to ensure that the axle head is square to the axle shaft.

8

Repeat the process for the other three axles, and then polish the axles to the desired finish.

Another important thing you can do to improve the performance of your car is to reduce the contact surface area between the wheel and the axle. We will reduce the contact surface area by cutting a groove in the axle. The part of the axle where the groove is cut never touches the wheel. This feature significantly improves the performance, especially when using liquid lubricants, with benefits when using dry lubricants as well.

This little speed tip also has a hidden advantage. The groove you cut in the axle becomes a secret trough where graphite gets stored. Later on during the race, some of that graphite will work its way out of the trough and be used to lubricate the inside of your wheels.

This speed tip is best accomplished with a drill press. However, if you are very careful and precise, you can do it with a power drill as well. The cuts you make in the axle must be absolutely straight. If you cut the groove on an angle, the wheel will tend to wobble as it turns.

2 Clamp the axle in your drill press or power drill. If using a drill press, line up the press table with the marks you made on the axle. Use the file to check that everything is aligned.

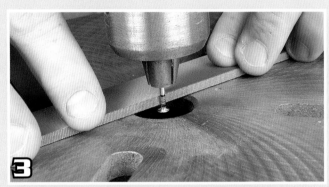

3 Using the flat file, gently cut a groove in the axle to a depth of about 1/16". Don't cut the groove too deep; it will weaken the axle.

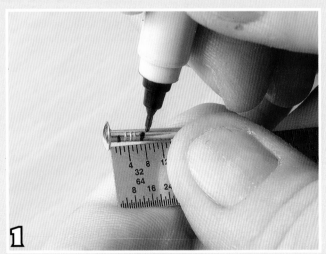

1 Place a dark mark on the axle 1/16" from the head. Place a second mark at 1/4".

4 Once you have finished polishing the axle according to the method on page 50, the final product should look like the one shown here.

Nathan T
Lancaster, PA

This car incorporates most of the tips from the Ultimate sections in this book. Notice the amount of wood that has been removed from the middle of the car, the extended wheelbase, the quick start design, and the H-shaped wheels. The steps for the Ultimate Axles and tungsten weights were also used for this car. For a gallery of other winning designs, see page 90.

Chapter 4
Wheel Preparation

To be competitive, you must prepare your wheels for success. The stock wheels that come straight out of the box leave a lot to be desired when it comes to building a winning car. In this chapter, we'll do a number of things to improve them, once again working within the three skill levels—Winning, Champion, and Ultimate.

The Winning Wheels

Official BSA wheels, straight out the box, sometimes have undesirable features, such as rather large divots or bubbles. They can also have other abnormalities that will reduce the speed of your car if left intact. To have a competitive car, you must remove these defective features. The wheels should also be lightly sanded to make them smooth.

The very basic steps of wheel preparation are outlined in this section. You would be surprised at how many people either don't bother with wheel preparation at all or end up doing it wrong and creating more problems in the process. Just follow the steps outlined below, and you will be in good shape. Remember to check your local rules and regulations before modifying wheels.

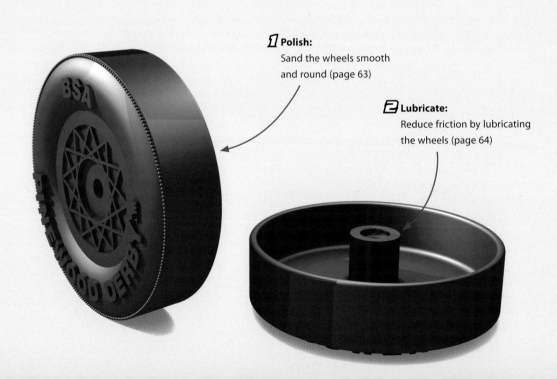

1 Polish:
Sand the wheels smooth and round (page 63)

2 Lubricate:
Reduce friction by lubricating the wheels (page 64)

Polish the Wheels

To get rid of the defects present on stock wheels, sand the wheels using a wheel mandrel. This handy tool is available at most BSA Scout Shops® or at *www.scoutstuff.org*. Do not attempt to sand the wheels without the use of a wheel mandrel! You will only create bigger problems for your car.

1

Attach a wheel to a mandrel, like this one from Derby Worx. Then, place the mandrel in an electric drill or a Dremel tool. If you choose to use a drill, clamp it in a vise, as shown here. Be careful not to tighten the vise so much that you damage your drill.

2

Attach a small sheet of 600-grit wet/dry sandpaper to a flat block of wood or sanding block. Apply water to the sandpaper to make it moist.

3

Turn on the drill or Dremel tool and allow the spinning wheel to gently rub against the surface of the sandpaper. Add more water to the paper when it starts to look dry. The plastic will become hot and actually melt unless you keep the sandpaper damp and maintain minimal pressure.

4

As you polish, be very careful not to apply too much pressure. Doing so will cause your wheel to become deformed.

5

Sand until the mold bubble is gone and the wheel looks smooth.

The Winning Wheels

No matter how much you polish your axles, there will still be friction as they make contact with the wheels. So, your next step in making a competitive car is to further reduce that friction by lubricating the wheels.

Your finished wheel should look like this one.

1

Select a lubricant that you will use for your wheels (see the Liquid vs. Dry Lubricant sidebar on page 65). I prefer to use graphite with molybdenum.

2

Rub a small amount of graphite onto a soft piece of cloth.

Using Graphite

When using graphite, it is important to wear a dust mask. Because graphite particles are very light, they can become suspended in the air, where they get inhaled. Graphite is also one of those substances that tends to get all over everything. To contain the graphite, find a suitable place outside or cover your working surface with newspaper, cloth, or plastic.

3

Polish the wheel with the graphite cloth until your wheel shines.
Note: Be sure to wear a dust mask when working with graphite.

Just Say "No" to White Teflon Powder

White Teflon powder was introduced as an alternative to graphite some years ago. Teflon works in frying pans, so it should work on Pinewood Derby wheels too, right? Absolutely not. Teflon powder doesn't work. It simply does not lubricate wheels as well as graphite. Stay away!

Liquid vs. Dry Lubricant

There are several types of lubricant to choose from. These are generally classified as either liquid or dry lubricants. Be sure to check with your local race official to find out what types of lubricant are allowed.

Liquid Lubricants

If your local rules allow the use of liquids, I would recommend NyOil II, a refined synthetic oil that works very well in Derby cars. NyOil II does outperform standard graphite with molybdenum by a significant margin when used correctly. However, be very careful. Liquids attract dust, dirt, and other particles. Mix any liquid with dirt and what are you going to get? Some type of paste that will act like cement in your axles. Exercise caution in the care of your wheels and axles to avoid letting dust or other particles get inside your wheel bore.

Dry Lubricants

There are actually several dry lubricants, but graphite with molybdenum is by far the most widely accepted and time-proven lubricant for Pinewood Derby cars. Molybdenum is a substance that acts like microscopic ball bearings.

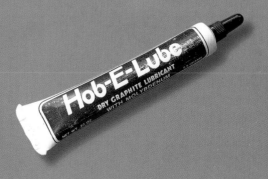

The Champion Wheels

The steps for the Champion wheels go far beyond the polishing and lubricating that were done for the Winning wheels. In this section, we'll look at some strategies for choosing the best wheels, we'll taper the wheel hub to reduce friction and gain speed, and we'll polish the inside rim so the car rides smoothly down the track.

Materials and Tools

- Electric drill or Dremel tool
- Wet/dry sandpaper, 400, 600, and 1000 grit, several sheets of each
- Small dish of water
- Vise
- Smooth-tooth file
- Wheel mandrel
- Flat block of wood or sanding block
- Lubricant of choice
- Soft cloth
- Safety goggles
- Dust mask

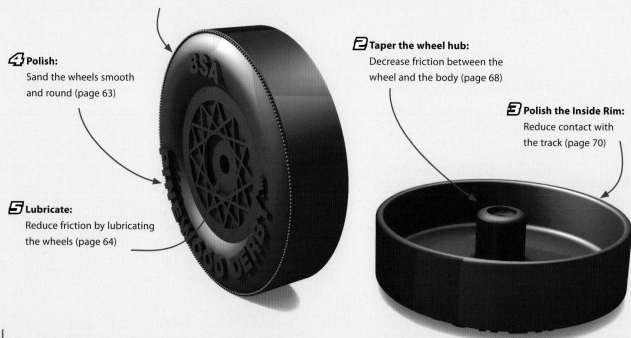

1 Use Premium Wheels:
Start with the smoothest and roundest Official BSA wheels (page 67)

4 Polish:
Sand the wheels smooth and round (page 63)

5 Lubricate:
Reduce friction by lubricating the wheels (page 64)

2 Taper the wheel hub:
Decrease friction between the wheel and the body (page 68)

3 Polish the Inside Rim:
Reduce contact with the track (page 70)

Use Premium Wheels

It is important to know that BSA wheels are manufactured using a 16-mold plastic injection process. Wheels are numbered 1 to 18, and mold numbers 6 and 7 are not included. The mold number associated with each wheel is stamped on the backside of each wheel. As of the printing of this book, wheels from die molds 2, 5, and 17 are better than other wheels. These wheels are usually defect free, have the truest shape, and have a larger wheel bore size. A larger bore size means that these wheels generate less friction with the axles.

You can purchase extra wheels at your local Scout Shop or from *www.scoutstuff.org*. Sort through the wheels and use those from mold numbers 2, 5, and 17 to build your car.

In past years, official BSA wheels produced from molds 9 and 13 were defective, as shown in the photo. As of the writing of this book, it appears the quality issues associated with molds 9 and 13 have been corrected. I examined thousands of wheels in 2005 and found very few defective wheels from these two mold numbers. I now feel confident in recommending both of these mold numbers for use in Pinewood Derby racing. However, please examine your wheels to be sure you did not get a set of defective 9 or 13 wheels that came from earlier stock. If you did, get a new set of wheels.

One of the largest sources of friction is the spot where the wheel hub rubs on the car body. Anything you can do to reduce friction at this location will help your car go faster. To this end, one of the very best things you can do is to taper the wheel hub using a smooth-tooth file. This little trick reduces the contact surface area by over 50%.

The Champion Wheels

1 Using a smooth file, create a tapered edge all the way around the wheel hub.

2 Sand the tapered edge with 400-grit wet/dry sandpaper. Then, sand using 600- and 1000-grit sandpaper.

3 Using a soft cloth, rub graphite into the tapered edge.

4 The final product should look like the wheel on the left.

Using the Derby Worx Pro Hub Tool

The Pro Hub Tool is designed to help you put a smooth, uniform bevel on the hub of each wheel. This handy tool will enable you to accomplish the task quickly and leave you with wheel hubs that look very professional.

1 The Pro Hub Tool comes with a protective plastic cover on both ends. Remove the protective cover. Notice that one end of the tool is square and the other end is cone-shaped.

2 Test fit each wheel by sliding it onto the tool pin. If the wheel bore is too small for the tool, use steady hand pressure and a twisting motion to work the wheel onto the tool. Remove the wheel and repeat 3 times for each wheel.

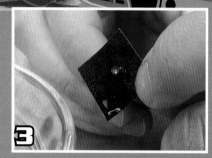

3 Dip the coarse-grit (150-grit) sandpaper that comes with the tool in a cup of tap water. Pierce the sandpaper (grit facing outward) with the square end of the tool, and slide the sandpaper until it makes contact with the body of the tool. To minimize the risk of a hand injury, leave the protective cap on the cone-shaped end of the tool.

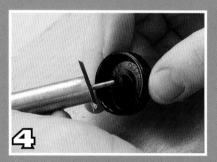

4 Place a wheel onto the square end of the tool, hub side first. With one hand, hold the tool and the sandpaper; with the other hand, press the wheel against the sandpaper. Rotate the wheel back and forth for about 10 seconds.

5 Replace the protective cap on the square end of the tool. Remove the protective cap from the cone-shaped end of the tool. Repeat Steps 3 and 4 using the coarse sandpaper on the cone-shaped end of the tool. Continue to repeat the process until the hub has a nice, defined cone shape.

6 Next, repeat Steps 3 and 4 using fine (220-grit) sandpaper on the cone-shaped end of the tool. When finished, the hub should look like the one shown here.

The Champion Wheels

This is one of those little tips that most people just forget about. As your car rolls down the track, the inside wheel rim will occasionally make contact with the center guide rail, so you need to make it as smooth as it can be. Polishing will reduce the friction.

1 Attach the wheel to a wheel mandrel and then spin the wheel using a power drill or a Dremel tool.

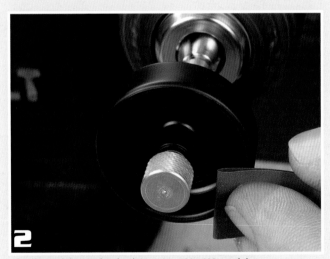

2 Gently sand the inside wheel rim using 400-, 600-, and then 1000-grit sandpaper.

3 Using a soft cloth, polish the inside rim with graphite. Remember to wear a dust mask and safety goggles whenever you work with graphite.

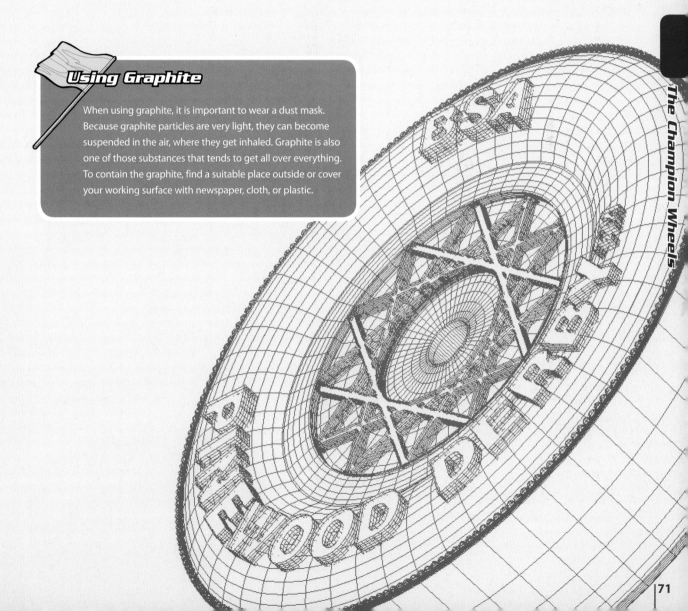

Using Graphite

When using graphite, it is important to wear a dust mask. Because graphite particles are very light, they can become suspended in the air, where they get inhaled. Graphite is also one of those substances that tends to get all over everything. To contain the graphite, find a suitable place outside or cover your working surface with newspaper, cloth, or plastic.

The Ultimate Wheels

The small steps that I'm illustrating in this section will really take your wheels to the next level. We'll discuss lathe-turned wheels and some different configurations that will greatly reduce friction between the track and the wheel. We'll also polish the last part of the wheel—the wheel bore—so that all parts your wheels are smooth.

Materials and Tools

- Electric drill or Dremel tool
- Wet/dry sandpaper, 400, 600, and 1000 grit, several sheets of each
- Small dish of water
- Vise
- Magnifying glass
- Smooth-tooth file
- Wheel mandrel
- Flat block of wood or sanding block
- Lubricant of choice
- Soft cloth
- Pipe cleaners with moderately stiff bristles
- Tube of whitening toothpaste
- Safety goggles
- Dust mask

1 Use Premium Wheels:
Start with the smoothest and roundest (page 67)

2 Lathe-turned Wheels:
Use smoother, rounder wheels (page 73)

3 Machine H or V Configurations:
Reduce the wheel surface area (page 74)

6 Polish the Wheels:
Sand them smooth and round (page 63)

4 Taper the wheel hub:
Decrease friction between the wheel and the body (page 68)

5 Polish the Inside Rim:
Reduce contact with the track (page 70)

7 Polish the Wheel Bore:
Less friction between the axles and wheels (page 75)

8 Lubricate the Wheels:
Reduce friction by lubricating the wheels (page 64)

Lathe-turned Wheels

One of the most powerful ways to improve the speed of your car is using lathe-turned wheels. As I mentioned before, stock wheels straight out of the box are not round! This fact can have a negative effect on how your car performs in competition, as shown by this experiment.

Find a basketball. Roll it around on the floor—it should roll smoothly. Now, let a little air out so that, if you sit or push on it, it will keep a portion of its shape. Now, roll the deformed ball across the floor. What happens? Does it roll straight and smoothly? Does it hiccup as it rolls? Does it wobble? Do you want your wheels to look like that as they roll down the track?

Lathe-turned wheels also have the advantage of weighing less than stock wheels. Simply put Less Weight = Less Inertia, and Less Inertia = Faster Starting Speed. Wheels with less inertia will begin turning sooner. In other words, at the top of the track, lightweight wheels will start rolling down the track before heavier wheels even begin moving.

Make no mistake, this is a very powerful speed tip! If your rules allow it and you can find a way to accomplish it, do it! It will make a huge difference in the speed of your car. However, there is one small catch to this tip. It requires the use of a machine lathe, which most people don't have lying around the house. A machine lathe is very expensive and requires skill to use.

Here's the bottom line:

- Don't attempt this step without a machine lathe. Some tip books suggest that you try to turn wheels with a mandrel and a power drill. Ignore this advice! Attempting to modify your wheels without a lathe is a recipe for disaster. You will almost certainly create problems that are worse than those you are trying to fix.

- Get a friend or a machine shop to turn your wheels. In my experience, most shops will be happy to work with you. However, it might cost a significant amount of money, and they might want to charge you their normal work rate to do the job.

The Ultimate Wheels

What's a Lathe?

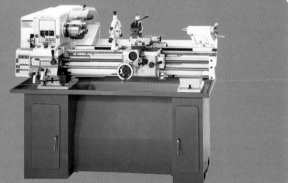

Here's a brief description of what a machine lathe does and how you can use it for perfecting your wheels. Once your wheel has been mounted on the machine lathe, the lathe spins your wheel while a blade removes all defects—leaving you with a smooth, perfectly round wheel. However, lathes are serious and potentially dangerous pieces of equipment. If you have access to one, adhere to all safety precautions and allow a professional to do the work. Do not attempt to turn your wheel on a wood lathe. And finally, under no circumstances should children be allowed to operate any lathe.

You can take lathe-turned wheels to an even higher level of performance by machining the wheel to an H or V configuration. Remember, friction is the enemy of speed, and light wheels are faster off the starting line than heavy wheels.

H and V configured wheels will reduce the amount of friction the wheel has with the track by reducing the surface area of the wheel. It also reduces the weight (or inertia) of the wheel. Be sure to check your local rules to see if they allow this modification.

H Configuration

In the H configuration, the only parts of the wheel that actually touch the track are the small rims on the outside of each wheel.

V Configuration

In the V configuration, the small point in the middle of the each wheel is the only part that actually touches the track.

Polish the Wheel Bore

You may think that you have polished every single location on your car that can possibly be polished. However, there is still one place left: the wheel bore.

I have actually looked down inside the wheel bore using some powerful magnifying equipment. Take my word for it—it's not a pretty sight! The inside of the wheel bore is anything but smooth. Some wheels even have long deep scratches down the bore.

You haven't spent tons of time polishing your axles only to insert them into a wheel bore that looks like sandpaper, have you?

I've spent years experimenting with methods for polishing the wheel bore, and I've tried every method I can find. Though some of the other methods being pushed on the Internet sound really good, this one is the best.

1 Secure the drill in a vise, being careful not to make it too tight.

2 Cut a 5" section of pipe cleaner and place it through one of the wheels.

Selecting Pipe Cleaner

Selecting a proper pipe cleaner is very important. The best ones can only be found at a tobacco store, but you can also try your local craft store. The right pipe cleaners have cotton fibers and fit snuggly into the wheel bore. Be sure that the kind you choose is soft but firm. Don't use the stiff, bristly type because they will actually damage your wheels.

3 Insert and secure the pipe cleaner in the drill. (This demonstration is continued on the next page.)

Polish the Wheel Bore

The Ultimate Wheels

4 Push the wheel toward the drill and coat the pipe cleaner with whitening toothpaste. I specified whitening toothpaste because this type of toothpaste contains a fine abrasive polish.

5 Hold the wheel between your fingers and turn on the drill using medium speed. Slowly move the wheel up and down the pipe cleaner as it spins. Polish each wheel for about 45 seconds. Don't overpolish; 45 seconds will do just fine.

6 Turn off the drill and remove the wheel from the pipe cleaner. Using warm water and a clean pipe cleaner, wash the wheel thoroughly. Be sure to remove all of the toothpaste from inside the wheel bore.

7 Using a third clean, dry pipe cleaner, dry the inside of the bore. Do not leave any water droplets inside the bore. If water droplets remain, they may leave residue behind as they dry. Repeat this process with all 4 wheels.

8 If you have a really good magnifying glass, you may be able to hold a wheel up to the light and look down inside the wheel bore. If you do, you should be able to see a very smooth polished surface—perfect for racing!

Corey R
Lancaster, PA

This car is an example of the proven wedge design. Also notice the use of the extended wheelbase and the H-shaped wheels. The steps for the Ultimate Axles were followed to complete this car. For a gallery of other winning designs, see page 90.

Chapter 5
Putting It All Together

Now that you have built your car, it's time to put everything together. This chapter will show you everything from attaching the wheels to preparing for the final weigh-in on race day. Please note that I have not included a list of tools and materials for these demonstrations because the items you'll need are things you've already been using or are common household items. Also included here are a few really good speed tips that a lot of people don't pay any attention to. Take it from me, they are powerful tips! Don't sell them short. They'll help turn your car into a rocket!

It's time to attach the wheels. Wheel attachment and spacing are extremely important. If you leave too much space between the wheel and car body, your wheels will shake and wobble as they turn. If you put them too close to the body, then the wheel hub will rub against the car almost the entire way down the track. So, we need to get the spacing right. The correct amount of space to leave between the wheel and the car body is ⅟₃₂". Before attaching the wheels, construct a wheel spacer to measure this.

1
Glue two or three business cards together (enough to equal a thickness of ⅟₃₂") with white glue. Use a tape measure or ruler to make sure your stack of cards is the proper thickness.

2
Cut a small slit approximately ⅛" wide by ¼" deep in the top of the cards with scissors.

3
Attach each wheel to the car (they should stay in place without any type of adhesive). Use the wheel spacer to control the amount of space left between the wheel and the car body. If you are using a raised front wheel, choose the wheel that spins the worst. To test the wheels, simply put them on an axle and spin them. The one that spins for the shortest time is the worst.

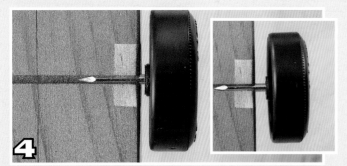

4
Your wheel should be spaced like the one in the photo. The inset shows an improperly spaced wheel.

5
Once all four wheels are attached to the car, place the car on a flat surface. The kitchen table or a countertop will work well. Get on your knees and look directly underneath the car. Each wheel should be resting flat and square on the table, unless you used a raised front wheel. If any of the wheels are not flat and square, adjust them until all four wheels are properly positioned.

Putting It All Together

Wheel Alignment

Proper wheel alignment is one of the most important aspects of a competitive car. If the wheels are not aligned, the car will not go straight. It will weave back and forth as it rolls down the track, bumping into the center guide rail over and over and slowing you down.

This is one of those speed tips that can really put you out in front of the pack because very few people understand proper wheel alignment and even fewer actually take the time to do it.

Your wheels should ride centered in the middle of their axles. As the wheel turns, you don't want it to move toward the body of the car because that will result in the wheel hub rubbing on the car body. It's not much better if the wheel moves outward toward the axle head, the wheel rubbing the axle head all the way down the track. Our first task will be to align the wheels so that they tend to stay centered in the middle of the axle as they turn.

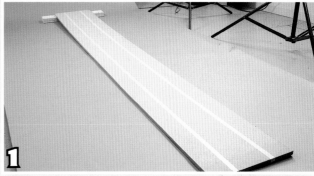

1 Make an inclined ramp that is 7' to 10' in length. A large sheet of plywood or a dining room table are two good possibilities. My family uses the dining room table with all three leaves installed to make it as long as possible. Use masking tape to mark a straight path down the middle of the ramp, as shown. The path should be about 6" wide.

2 Put some pillows at the bottom of the ramp to stop the car.

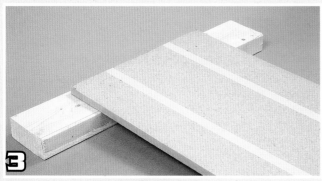

3 Prop the ramp up on a very slight angle.

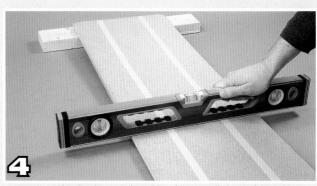

4 Use a 3' or 4' level to be sure the table is horizontally level. If the table is tilted right or left, the car will tend to roll that direction, and you will not be able to accurately align your wheels.

5 Position the car at the top of the ramp. Make sure each wheel is centered in the middle of its associated axle. Let the car start rolling very slowly down the ramp.

6 Allow the car to roll 2 or 3 feet, taking note of whether each wheel tends to move inward toward the car body, outward toward the axle head, or whether it just rolls in the middle of the axle. Repeat this process until you are confident that you understand how each wheel is moving.

7 If a wheel moves inward toward the car body, the axle needs to be adjusted up. If a wheel moves outward toward the axle head, the axle needs to be adjusted down.

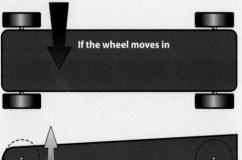

If the wheel moves in

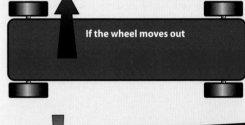

If the wheel moves out

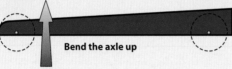

Bend the axle up

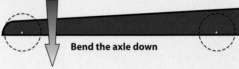

Bend the axle down

8 Adjusting the axles is the most important part of this process. You must do it correctly. The idea is to slightly bend the axle so that it angles up or down as needed without bending the portion of the axle where the wheel rides. Before removing the axle from the car body, place a mark on the axle at exactly 12 o'clock. This will ensure that you always know the original orientation of the axle. Also place a mark at the exact location where the axle meets the car body.

9 Carefully remove the axle from the car body. Be very careful not to damage the wheels in the process.

10

Clamp the axle in a vise or an axle press. If the axle needs to be adjusted up, make sure the 12 o'clock mark is away from you, as shown. If the axle needs to be adjusted down, make sure the 12 o'clock mark is toward you. The mark you made in Step 8 should be located slightly above the edge of the vise.

11

Place a flat, regular screwdriver squarely at the base of the axle, slightly below the mark you made in Step 8. Gently tap on the screwdriver with a small hammer or rubber mallet. This will create a slight bend at the marked location. You don't want to bend the axle too much. Small adjustments are the key to the process.

12

Remove the axle from the vise and reattach the axle and wheel to the car body. Be sure the 12 o'clock mark is pointing straight up.

13

Place the car on the ramp and allow the car to roll down the ramp 2 or 3 feet. If the wheel stays pretty much in the center of the axle, then you are finished with up/down adjustments for that axle. If the axle continues to move in toward the body or out toward the axle head, you need to repeat the process until the wheel remains in the center of the axle. This may take some time. There have been occasions when we repeated this process 7 or 8 times before everything was perfect.

Putting It All Together

Wheel Canting or Angling

This method has received quite a bit of press on numerous websites and in books. The idea is to angle the axles upward so that the wheels are forced toward the axle head. Supporters of this idea suggest that it's better to have the wheel always rubbing on the axle head than it is to ever let it touch the car body. They also suggest that canted wheels go faster because less of the wheel actually touches the track.

While it is true that contact with the axle head is better than contact with the car body and that having less wheel contact with the track is advantageous, there are two big issues with wheel canting. First, angling the axle causes the wheel to ride in an orientation that is not parallel to the axle. This results in tremendous binding and friction between the wheel and the axle. Second, although it is preferable to have contact with the axle head, it is not preferable to have contact with the axle head 100% of the time. The moral of the story is that canted wheels create more friction than non-canted wheels. And remember, friction is the enemy of speed.

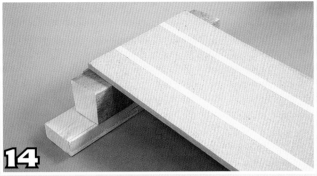

14

Once all four wheels ride in the middle of the axle consistently, it's time to make your car go straight. Elevate your ramp so that, when released, your car will roll down the ramp at a fairly good speed. Be sure you have pillows at the end of the ramp.

15

Place the car at the top of the ramp. Make sure it is lined up straight in the middle of the taped path.

16

Release the car and let it roll all the way down the track. Did it stay within the taped path? Did it roll to the right or to the left? If it didn't roll straight down the middle of the path, then you need to make some adjustments to the front axles.

17

If you only have three wheels touching the track, you will need to adjust only the front wheel that is touching. If you have both wheels on the track, then you will need to identify which wheel is dominant, or which wheel is carrying most of the weight. If you only have three wheels on the track, move on to Step 19. Otherwise, follow the instructions to find the weight-bearing, dominant front wheel.

18

Place the car on a flat, level surface. Then, place your index finger directly on the front center of the car. Gently press down. As you press down, you will feel the car slightly shift toward one of the front wheels. The dominant wheel is the wheel in the opposite direction of the shift. This is the wheel that carries the weight.

**If the car moves left,
bend the axle forward**

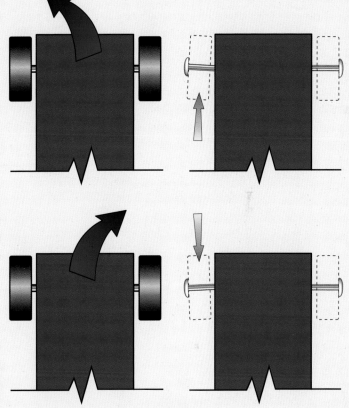

 Check to be sure the dominant axle head has a mark located at 12 o'clock.

Remove the axle and repeat the adjustment process described in Steps 7 through 12.

However, this time you will be adjusting the axle left and right. If your car turns to the left as it rolls, then adjust the axle right. If the car rolls right, then adjust the axle left.

**If the car moves right,
bend the axle backward**

20 Repeat this process until your car rolls straight down the ramp consistently 5 or 6 times. When all four axles are adjusted to perfection, you are ready to glue them in place.

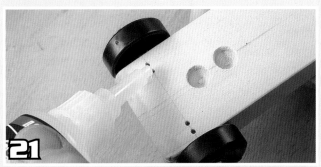

21 Turn your car over on its back and apply one or two drops of CA glue in the center of each axle. Allow the glue to dry overnight before turning your car back over. If using the standard axle slots, do not use CA glue. Because it is very thin and runny, CA glue tends to run down the axle and get on the wheel hub. Instead, try some type of quick-drying cement or epoxy.

Believe it or not, the method you use to apply graphite can also make a difference in the speed of your car. In order for the graphite to be effective, it needs to be placed inside the wheel bore. However, once the wheel has been attached to the car, inserting graphite into the small space between the wheel and car body can be somewhat of a problem. Most people will tell you to squirt the graphite into the wheel using the pointed tip on the end of the tube. In my experience, this doesn't work very well. Precious little graphite actually makes it down inside the wheel where it belongs. The very best method I've found is to use a small paintbrush to apply the graphite (those small brushes that come in watercolor kits work great). Remember to wear a dust mask and safety goggles whenever you are working with graphite.

1 Take your tube of graphite and empty it into a small dish. Dip the brush into the graphite, and then gently transfer the graphite from the brush to the car.

2 As you apply the graphite, gently shake or tap each wheel. This will help move the graphite down inside the wheel bore.

Breaking Them In

I love this speed tip! If you have ever been to a Pinewood Derby, you may have noticed that a lot of cars will not do very well their first or second time down the track. However, by the end of the Derby, those same cars will be running better and will sometimes even beat cars they lost to in the first or second heat.

This isn't just your imagination playing tricks on you. It really happens. As a general rule, your car will get a little faster each time it goes down the track. Each time you race, the wheels and axles wear into each other a little bit more. They get smoother and smoother. Your car will be faster the twentieth time it races than it was the first time.

So, hold on to your hat when you hear this speed tip: Race your car 50 or 100 times before you ever show up at the Pinewood Derby! How do you accomplish this without a track? It's simple. The method I'll show here requires a Dremel tool with a large felt disk attachment. If you don't have a Dremel tool, don't worry. You can use your index finger instead. This is another step that your child can perform almost all on his own.

2 Turn the Dremel tool on high speed. Hold the car level and gently touch the spinning felt disk to one of the wheels. Once the wheel is spinning, move the Dremel tool away from the wheel. If you hold the felt disk to the wheel continuously, you can damage it. As you see the wheel start to slow down slightly, touch the disk to it again. Repeat this process for about 5 minutes per wheel. If you are using your index finger to do this, repeat the process for about 15 minutes per wheel. You may want to put in a good movie.

1 Go outside or to the garage (or choose another suitable area in case the graphite makes a mess) and apply graphite to each wheel. Spin the wheels a little bit. Remember to wear goggles and a dust mask.

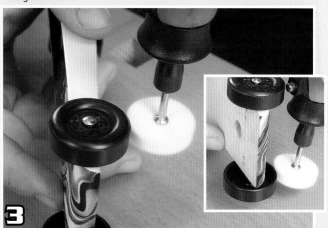

3 Turn the car on its side and spin the wheels. Turn the car over on the other side and spin them. Every minute or so apply more graphite. Spin each wheel for about 5 minutes on high speed. If you are using your index finger, periodically apply more graphite (after about every 5 or 10 minutes), and continue spinning your wheels until the movie is over. When you are finished, your car will have run a good 100 races or more. It will be a fine-tuned racing machine!

Basically, the 20-second test is the final test you will give your car to be sure everything is correct and ready to go. After the wheels have been lubricated, aligned, and broken in, spin each wheel with your finger as fast as you can. Time how long each wheel spins. Repeat this procedure 5 or 6 times for each wheel.

If everything is right, your wheels should spin for at least 20 seconds. If you find one wheel that just doesn't measure up, try adding more graphite. Spin that wheel again for a minute or so to break it in. When finished, try the 20-second test again. If the wheel now spins for 20 seconds, great! If it's still slow, I would recommend that you remove the wheel/axle pair and replace them with another set. If you end up replacing a wheel, then you will need to re-align your car.

In my experience, if a wheel will not spin for at least 20 seconds, then something is wrong. It's not performing at the level it should if you want to have a winning car. If you have done everything correctly, your wheels might spin for longer, up to 34 seconds.

Test that each wheel spins for at least 20 seconds. Doing so will ensure that everything is correct and ready to go.

Putting It All Together

The Final Weighing

I've said this before, and I'll say it again. The heavier your car is, the faster it will go. You need to have every fraction of a gram's worth of weight the judges will allow. Show up at the check-in station with your car slightly under weight, say 4.95 ounces. Place your car on the scale, and then slowly add small bits of weight to the scale. Continue to add small bits of weight until the scale reads *more* than 5.0 ounces. Then, remove the last bit of weight you added.

There is an important principle at work here. On a scale that measures to the nearest 0.1 ounces, a car weighing 4.96 ounces and a car weighing 5.04 ounces both register as 5.0 on the scale. So, make sure you have the maximum amount of weight allowed.

We like to use an adjustable weight, such as tungsten putty, small split shot fishing weights, modelling clay, or brass screws, for this part of the process. This type of weight fits easily into one of the small holes we have previously drilled in the bottom of the car (if you use tungsten putty, be sure to wash your hands and follow safety precautions during and after its use).

The other reason this speed tip will help propel you to the front of the pack is this: A lot of people show up with cars that weigh too much. In order to reduce the weight of their car, they end up cutting wood off the car or drilling holes in the bottom. If wood chips or sawdust get into your wheels, you can kiss your chances of winning goodbye. Why would anybody spend so much time preparing his wheels and axles and then blow sawdust on them ten minutes before the race? Show up a little under weight, and then add to it.

A note to parents: At most weigh-ins, you will see parents lining up to place their "child's" car on the scale. Don't do this. This evening belongs to your child. Allow him to place his own car on the scale. I actually practice this activity with my sons at home. We put our scale on the table, and I allow them to practice carefully placing the car on the scale, upside down so that it won't roll off. They practice putting small bits of tungsten putty on the scale until it reads 5.0 ounces. When we show up at weigh-in, my sons know exactly what to do and how to do it.

In closing, please remember that the most important goal of the Pinewood Derby is to have fun! Have fun by being safe, being honest, and showing good sportsmanship. Even though we have spent a lot of time talking about principles that will help you win the Pinewood Derby, don't forget that your Derby experience is only a success if you enjoy the ride.

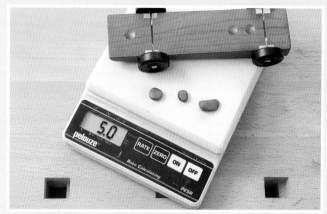

An adjustable weight, such as tungsten putty, is perfect for getting the weight to exactly five ounces on race day.

Small bits of tungsten putty have been added to the holes as necessary.

Putting It All Together

Tyler V
Fenelton, PA

Julie S
Austin, TX

Cole W
Orange, CA

Cole W
Orange, CA

Damien C
Farmer City, IL

Damien C
Farmer City, IL

Ryan U
Louisburg, KS

Kaitlyn U
Louisburg, KS

Collen R
Lancaster, PA

Corey R
Lancaster, PA

Tyler P
Springwater, NY

Nathan T
Lancaster, PA

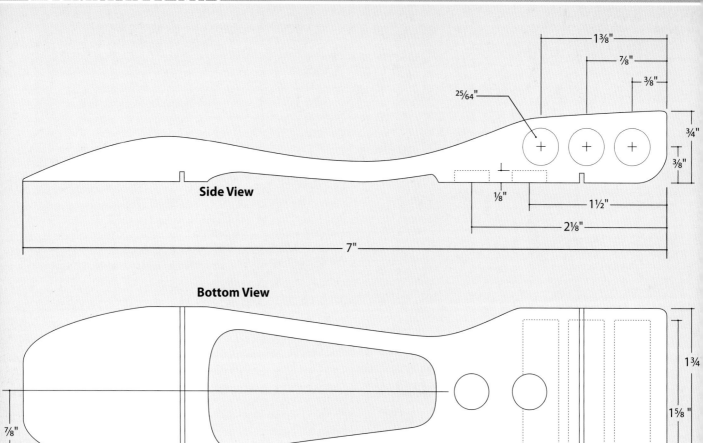

Side View

1³⁄₈"
⁷⁄₈"
³⁄₈"
25⁄₆₄"
³⁄₄"
³⁄₈"
¹⁄₈"
1½"
2¹⁄₈"
7"

Bottom View

1¾
1⁵⁄₈"
⁷⁄₈"

Cutting Guides

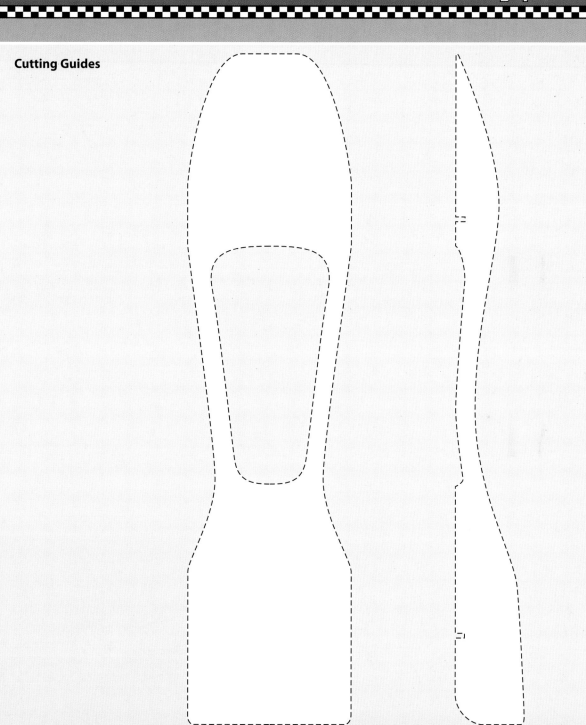

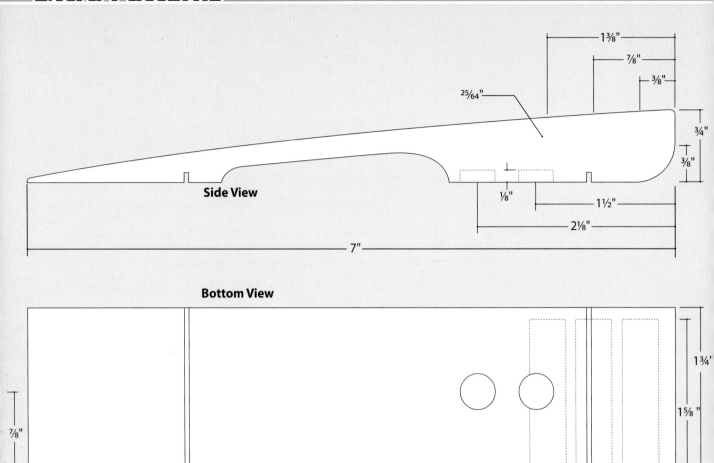

Side View

1³⁄₈"
⁷⁄₈"
³⁄₈"
²⁵⁄₆₄"
³⁄₄"
³⁄₈"
¹⁄₈"
1½"
2¹⁄₈"
7"

Bottom View

1¾'
1⁵⁄₈"
⁷⁄₈"

Cutting Guides

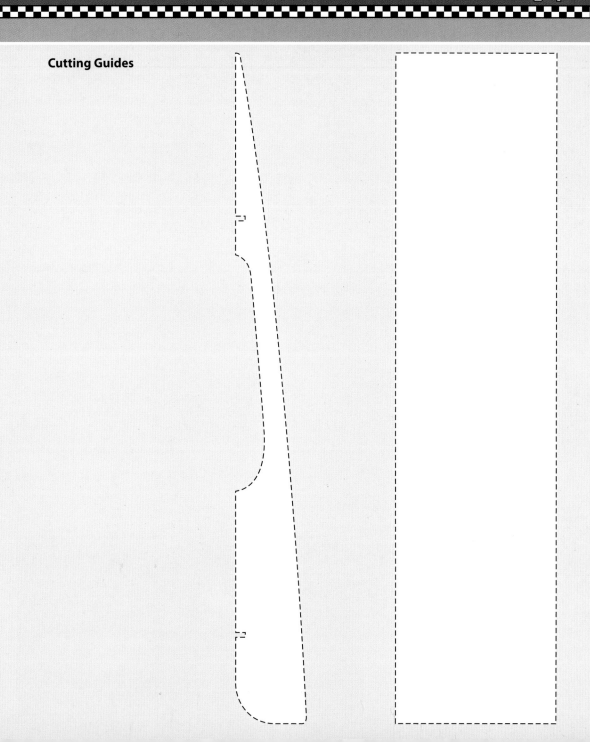

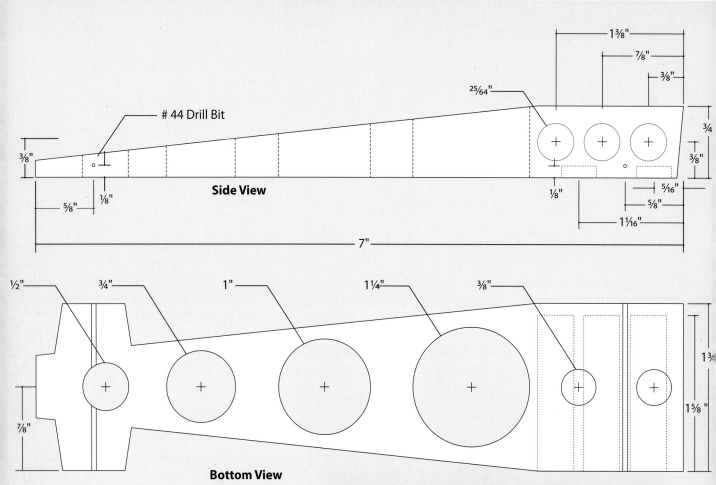

1³⁄₈"

⁷⁄₈"

³⁄₈"

²⁵⁄₆₄"

44 Drill Bit

³⁄₄

³⁄₈"

³⁄₈"

Side View

⁵⁄₁₆"

⁵⁄₈"

¹⁄₈"

¹⁄₈"

⁵⁄₈"

1¹⁄₁₆"

7"

¹⁄₂"   ³⁄₄"   1"   1¹⁄₄"   ³⁄₈"

1³⁄

⁷⁄₈"

1⁵⁄₈"

Bottom View

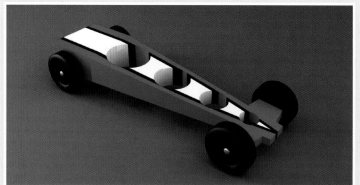

Cutting Guides

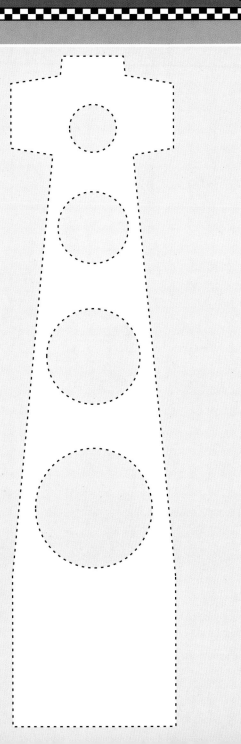

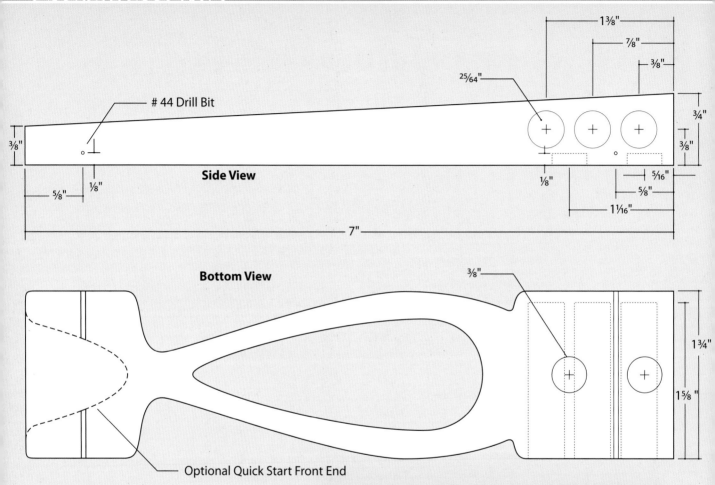

Side View

44 Drill Bit

25/64"

1 3/8"

7/8"

3/8"

3/4"

3/8"

3/8"

5/8"

1/8"

1/8"

5/16"

5/8"

1 1/16"

7"

Bottom View

3/8"

1 3/4"

1 5/8"

Optional Quick Start Front End

Cutting Guides

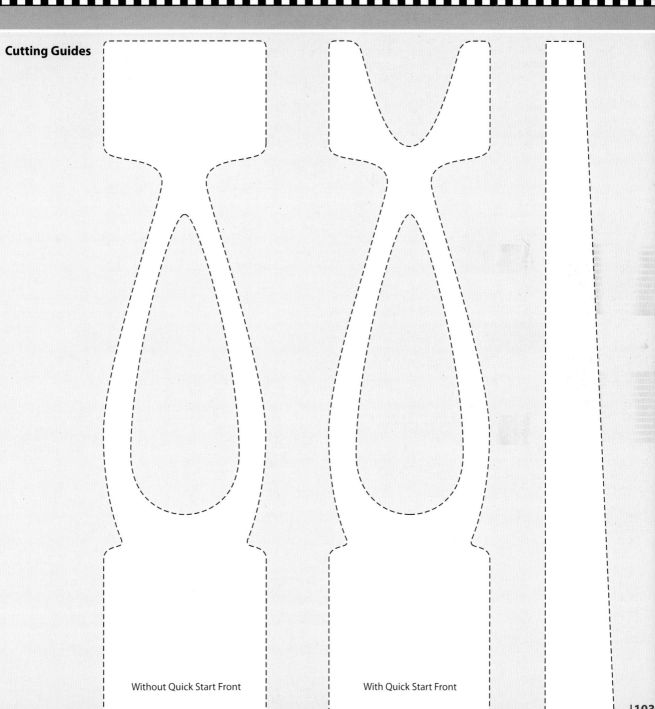

Without Quick Start Front

With Quick Start Front

Standard Wheelbase

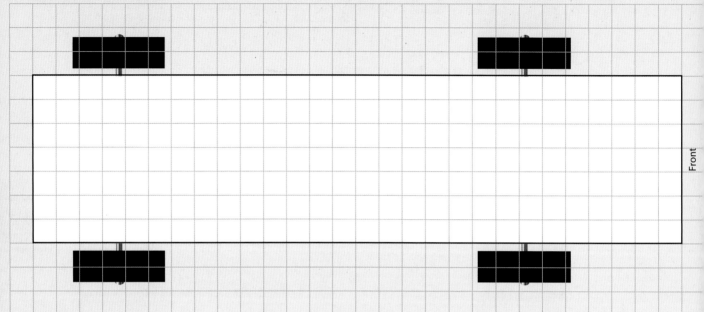

Top View

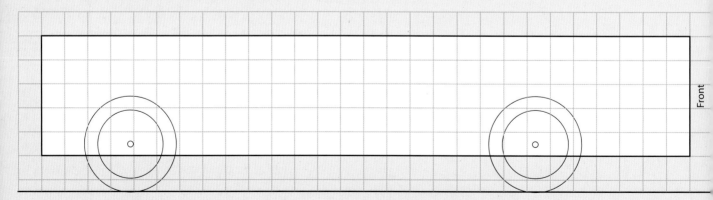

Side View

Extended Wheelbase

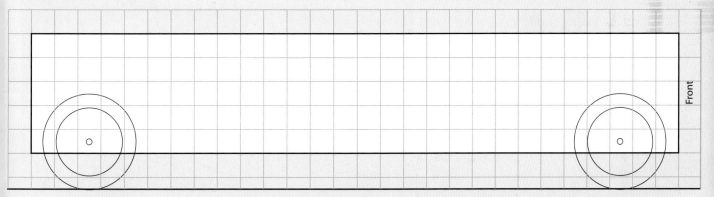

Top View

Front

Side View

Front

For more secrets and information, visit David Meade's website at **www.pinederbyinfo.com**.

Resources

Official Pinewood Derby Supplies

Boy Scouts of America
P.O. Box 7143
Charlotte, NC 28241-7143
1-800-323-0736
www.scoutstuff.org

Book on Pinewood Derby History

Don Murphy
P.O. Box 3881-B
Torrance, CA 90510
(310) 320-4343
www.pinewoodderbystory.com

Rotary/Shaper Tools

Dremel
4915 21st Street
Racine, WI 53406
1-800-437-3635
www.dremel.com

Derby Timers

Micro Wizard Derby Timers
10007 Old Union Road
Union, KY 41091
1-888-693-3729
www.microwizard.com

Derby Tracks

Piantedosi Oars
P.O. Box 643
West Acton, MA 01720
(978) 263-1814
www.pinewoodderbytrack.com

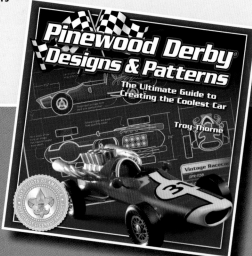

Build the Coolest Car in the Pinewood Derby!

Design a cool-looking car that's quick out of the gate! *Pinewood Derby Designs & Patterns*, licensed by the Boy Scouts of America and written by Troy Thorne, a woodworker, artist, and Derby-winning dad, contains all the easy-to-follow instructions and amazing patterns to make your child's Derby car the coolest in the race. Troy Thorne shares his expert tips, techniques, and secrets for every step of the process, right down to the final decals. Inside *Pinewood Derby Designs & Patterns*, you'll discover:

- Easy-to-follow steps for building the High-Wing Racer, Stock Car, and Vintage Racecar

- How to distribute weight for maximum speed

- 34 jaw-dropping patterns and designs

- Techniques for creating custom decals and applying a high-quality finish

- Helpful tips for prepping the car's wheels, axles, and weight for the race

With *Pinewood Derby Designs & Patterns*, parents and scouts of any skill level can work together to build a great looking, prize-winning car while building memories that last a lifetime!

Look for *Pinewood Derby Designs & Patterns* at your local bookstore or contact the Boy Scouts of America at 1-800-323-0736 or *www.scoutstuff.org*.